Research Reports in Physics

Research Reports in Physics

Nuclear Structure of the Zirconium Region
Editors: J. Eberth, R. A. Meyer, and K. Sistemich

Ecodynamics Contributions to Theoretical Ecology
Editors: W. Wolff, C.-J. Soeder, and F. R. Drepper

Nonlinear Waves 1 Dynamics and Evolution
Editors: A. V. Gaponov-Grekhov, M. I. Rabinovich, and J. Engelbrecht

Nonlinear Waves 2 Dynamics and Evolution
Editors: A. V. Gaponov-Grekhov, M. I. Rabinovich, and J. Engelbrecht

Nuclear Astrophysics
Editors: M. Lozano, M. I. Gallardo, and J. M. Arias

Optimized LCAO Method and the Electronic Structure of Extended Systems By H. Eschrig

Helmut Eschrig

Optimized LCAO Method

and the Electronic Structure of Extended Systems

With 80 Figures

Springer-Verlag Berlin Heidelberg GmbH

Dr. Helmut Eschrig
Zentralinstitut für Festkörperphysik und Werkstofforschung
der Akademie der Wissenschaften der DDR
Helmholtzstraße 20, 8027 Dresden, GDR

Sole distribution rights for the non-socialist countries
Springer-Verlag Berlin Heidelberg New York
London Paris Tokyo

ISBN 978-3-662-02564-2 ISBN 978-3-662-02562-8 (eBook)
DOI 10.1007/978-3-662-02562-8

Originally published by Springer-Verlag Berlin Heidelberg New York 1989
Softcover reprint of the hardcover 1st edition 1989

PREFACE

Since the late seventies, the theory of the electronic structure of ordered crystalline solids has gone through a period of dramatic development, triggered by the use of powerful computers to apply such tools of many-body theory as the density functional formalism and Green's function methods. Precise and effective numerical procedures for solving the one-particle Schrödinger equation for given arrangements of atoms are the key to the application of modern electronic structure theory. There are quite a few almost equally successful approaches here, such as KKR, APW, their linearized versions, norm-conserving pseudo-potentials and others, but the final results at least are best represented in the common language of the linear combinations of atomic orbitals (LCAO) [100]. Especially the interpretation of modern electronic structure theory in chemical terms [57] made LCAO representations very popular. A non-orthogonal local orbital representation has been used with remarkable success to deal with electron-phonon interaction and with phonon anomalies in metallic solids [117].

Having taken part in the research activities of the Dresden group working on metal theory for the past two decades [120], the present author developed the numerics for a first-principles self-consistent field LCAO method based on a special optimization procedure for the basis states, and applied it successfully to many electronic structure problems. The advantage of a complete LCAO treatment compared with LCAO interpolation schemes is obvious: it provides the user with the explicit forms of the wavefunctions. The crucial problem for variational procedures like LCAO is the proper choice of basis functions. In the version presented here the ordinary LCAO treatment is preceded by another variational parameter adaption depending only on the structure type of the lattice, not on the potential.

The present book gives a full description of the author's version of the SCF-LCAO method including all numerical details needed to apply it, and considers the main physical ideas closely related to it. Also it provides the reader with complete data on SCF-LCAO bandstructure results for all element metals up to atomic number 30. Extensions to multi-component systems, spin-polarized and relativistic cases, disordered solids, and clusters, which do exist, will be presented in subsequent volumes.

Besides the already mentioned Dresden group, the author's approach to electronic structure theory has directly or indirectly been guided by many people. It is a pleasure for him to take this opportunity to thank all of them, particularly Profs. P. Ziesche, G. Lehmann, Y.M. Kagan, M.I. Kaganov, W.A. Harrison, and W. Weber. The work presented here was done in the Zentralinstitut für Festkörperphysik und Werkstofforschung Dresden of the Academy of Sciences of the GDR. The numerical calculations were supported by the staff of the computer station of the ZfK Rossendorf. Last but not least, thanks go to my son Matthias who typed the complicated text with great patience.

Dresden, April 1987 Helmut Eschrig

CONTENT

Notation:
Bold-face type is used for vectors $\mathbf{R} = (R_1, R_2, R_3)$ or matrices $\mathbf{A} = [[A_{ij}]]$. The absolute value (Euklidean norm) of a vector $\mathbf{R}$ is R. A roof ^ on a letter in ordinary type, $\hat{H}$, denotes an operator, on a bold-face type, $\hat{\mathbf{r}}$, it means the unit vector in the direction of $\mathbf{r}$. For technical reasons, the centre Γ of the Brillouin zone (B.Z.) is denoted by 0 in bandstructure plots. All other notations are explained in the text.

1. INTRODUCTION

The present treatise deals basically with a special method of solving quantum mechanical one-particle equations for systems composed of atoms or ions. This introduction sketches the ways those equations appear in modern theory. The ideas and formal developments are presented only inasmuch as they are related to the basic content of this volume; the following introduction cannot be regarded as a comprehensive survey over the theories sketched. Throughout the whole text of this booklet a restraint to the non-relativistic description is observed in order to keep the argumentation as well as the formal developments within a definite notion. Such a restriction is neither inherent the theoretical constructions nor the technical tools; practically everything treated in this volume may be generalized to the relativistic cases of inner shell motion in heavy atoms, spin-orbit coupling and so on.

1.1. The adiabatic approximation

The non-relativistic motion of a system of nuclei a with masses M_a and charges A_a (in units of the electrostatic proton charge $|e|$, the electrostatic charge differs from the usual electric charge by a factor $1/\sqrt{4\pi\varepsilon_0}$) and of electrons with mass m may at least formally be obtained from the Hamiltonian[1)]

$$\hat{H}_{tot} = \sum_{a}^{N} \frac{\nabla_a^2}{2M_a} + \sum_{a \neq a'} \frac{A_a A_{a'}}{|\mathbf{R}_a - \mathbf{R}_{a'}|} + \hat{H}, \tag{1.1}$$

$$\hat{H} = \sum_{i}^{N_e} \frac{\nabla_i^2}{2} + \sum_{i \neq i'} \frac{1}{|\mathbf{r}_i - \mathbf{r}_{i'}|} - \sum_{i,a} \frac{A_a}{|\mathbf{r}_i - \mathbf{R}_a|} = \hat{T} + \hat{W} + \hat{V}. \tag{1.2}$$

Here,

$$\mathbf{R} = (\mathbf{R}_1, \ldots, \mathbf{R}_a, \ldots, \mathbf{R}_N) \tag{1.3}$$

are the coordinates of the nuclei and $\mathbf{r}_i$, $i = 1, \ldots, N_e$ are those of the electrons. Due to the large mass difference the electrons move much faster than the nuclei. If the latter were fixed at positions $\mathbf{R}$, the

1) Atomic units (a.u.) $m = \hbar = |e| = 1$ are used throughout. The energy unit is 1 a.u. = 1 Hartree = 2 Rydberg.

electronic ground-state $\Psi(x_i;\mathbf{R})$ could be calculated from the Hamiltonian $\hat{H}$. In the argument of Ψ,

$$x_i = (\mathbf{r}_i, s_i) \tag{1.4}$$

contains the electron coordinates and spin variables; though the Hamiltonian $\hat{H}$ does not contain the electron spins, its ground-state may depend on them via the Pauli principle demanding $\Psi(x_i;\mathbf{R})$ to be totally antisymmetric in the variables x_i and thus resulting in the exchange interaction of the electrons. The adiabatic theorem of Gell-Mann and Low [45], applied to our situation, states that the electronic subsystem would remain in its ground-state all the time if the nuclei were infinitely slowly moved. The electronic ground-state energy $E(\mathbf{R})$ would then be obtained from the equation

$$\hat{H}\,\Psi(x_i;\mathbf{R}) = \Psi(x_i;\mathbf{R})E(\mathbf{R}) \tag{1.5}$$

parametrically depending on $\mathbf{R}$. Integrating over all but one electronic coordinates and summing over all spin variables the electronic ground-state charge density $-\varrho(\mathbf{r};\mathbf{R})$ (ϱ being the number density in our units) is obtained :

$$\varrho(\mathbf{r};\mathbf{R}) = \sum_{s}\sum_{x_2}\dots\sum_{x_{N_e}} |\Psi((\mathbf{r},s),x_2,\dots,x_{N_e};\mathbf{R})|^2, \tag{1.6}$$

where the abbreviation

$$\sum_{x_i} \equiv \sum_{s_i=-1/2}^{1/2}\int d^3r_i$$

is used. A great part of what follows just deals with the computation of $\varrho(\mathbf{r};\mathbf{R})$ and $E(\mathbf{R})$.

Within this frame the motion of the nuclei would be obtained from the Hamiltonian

$$\hat{H}_n = \sum_{a}^{N}\frac{\nabla_a^2}{2M_a} + \sum_{a\neq a'}\frac{A_a A_{a'}}{|\mathbf{R}_a-\mathbf{R}_{a'}|} + E(\mathbf{R}). \tag{1.7}$$

After eliminating the trivial motion of the centre of mass of the nuclei, (1.7) yields vibrations around the minimum of

$$\boxed{U(\mathbf{R}) = \sum_{a\neq a'}\frac{A_a A_{a'}}{|\mathbf{R}_a-\mathbf{R}_{a'}|} + E(\mathbf{R}).} \tag{1.8}$$

Often the minimum configuration $\mathbf{R} = \mathbf{R}_o$ and the corresponding potential value $U(\mathbf{R}_o)$ are of sufficient interest by themselves implying structural information and binding energies, the latter as differences between $U(\mathbf{R}_o)$ and the sum of the corresponding values for the separated constituents of the considered system.

In reality the nuclei move with a finite velocity, yet the outlined frame may be well founded as the zeroth order approximation of the Born-Oppenheimer perturbation theory [11,10,94]. The non-adiabatic coupling of the electronic and nuclear motion appears in the higher perturbation orders with respect to the small parameter $(m/M_a)^{1/2}$. There are basically two cases where the adiabatic approximation breaks down. One is if the ground-state of the electronic system is almost degenerate so that the lowest electronic excitation energy is comparable with the vibrational energies. This happens e.g. in Jahn-Teller systems [64,63,23] or in the predissociation of molecules. In the other case the lowest electronic excitation energy may be as large as one Rydberg, but due to the lack of screening the electron-nuclei interaction matrix element diverges. This leads e.g. to the polaron mode in insulators [56].

In metals the electronic excitation spectrum has no gap. Consequently there are always electronic excitations near the Fermi surface subject to large non-adiabatic corrections. However, the phase space of these states is so small that the corrections do not substantially contribute to total energies [13] save again for metallic Jahn-Teller systems [42,43,29,118].

1.2. Density functional theory (DFT)

The adiabatic approximation is only a first step simplifying the problem. The resulting equation (1.5), as an eigenvalue problem for a partial differential equation in 10 to 10^{23} variables, depending on the considered system, is still far from being subject to direct numerical calculations. There is, however, an indirect way attacing this task which starts from the basic theorem of Hohenberg and Kohn [60] stating that there is a one-to-one correspondence between the potentials $\hat{V}$ of (1.2) and the ground-state electron densities (1.6) for a given number N_e, provided $\hat{H}$ has a non-degenerate ground-state Ψ. Hence, $\hat{V}$ may be regarded as a unique functional of the electron density $\varrho(\mathbf{r})$. On the other hand, as the solution of (1.5), the ground-state $\Psi(x_i)$ and consequently any ground-state property, particularly the ground-state energy E, are unique functionals of $\hat{V}$. This implies that those quantities are also unique functionals of $\varrho(\mathbf{r})$:

$$\Psi = \Psi[\varrho], \quad E = E[\varrho] = (\Psi[\varrho], \hat{H}[\varrho]\,\Psi[\varrho]). \tag{1.9}$$

The notation $\hat{H}[\varrho]$ means that we have expressed $\hat{V}$ in terms of ϱ. Defining the Hohenberg-Kohn functional

$$F[\varrho] = (\Psi[\varrho], [\hat{T} + \hat{W}]\, \Psi[\varrho]) \tag{1.10}$$

and writing $\hat{V}$ as

$$\hat{V} = \sum_{i=1}^{N_e} v(\mathbf{r}_i), \tag{1.11}$$

where again $v = v[\varrho]$ holds, we may write

$$E[\varrho] = F[\varrho] + \int d^3r \varrho(\mathbf{r}) v(\mathbf{r}). \tag{1.12}$$

Now, recalling that the ground-state energy $E[v]$ is the minimum of $(\Psi, \hat{H}\,\Psi)$ with respect to variations of Ψ, it is easy to find

$$E[v] = \min\{\, F[\varrho] + \int d^3r \varrho v \mid \int d^3r \varrho = N_e \,\} \tag{1.13}$$

where $\{\, A \mid C \,\}$ means the set of expressions A fulfilling the constraint C. This is the famous variational principle of Hohenberg and Kohn. Note that in (1.13), contrary to (1.12), v is fixed when ϱ varies.

A weak point of this variational principle is that neither the set of potentials v having non-degenerate ground-states nor the set of densities ϱ to vary within are explicitly known. For this reason, various alternatives to (1.13) have been formulated (see [90] for a survay on the level of mathematical physics), but most applications use (1.13) in a pragmatic manner. If the admissible ϱ are supposed to be dense in some functional space and there the functional derivative of $F[\varrho]$ is supposed to exist (again a point open for discussion; for finite systems see [30], the case of an extended system is considered below), then from (1.13) follows

$$\frac{\delta}{\delta\varrho} \{\, F[\varrho] + \int d^3r \varrho v - \mu \int d^3r \varrho \,\} = 0,$$

that is

$$\frac{\delta F}{\delta\varrho} + v = \mu. \tag{1.14}$$

Having added the constraint of (1.13) with a Lagrangean multiplier μ the density ϱ is already freely varied in (1.14).

The main drawback of the approach is that the functional $F[\varrho]$ is unknown. Due to the well developed theory of the homogeneous electron liquid it is best known for homogeneous densities ϱ=const. This could be used for a Thomas-Fermi like approximation, but like in Thomas-Fermi theory one would be left with a very crude approximation of the kinetic energy. Just this point led Kohn and Sham [75] to a very ingenious trick. First F is split according to

$$F[\varrho] = T[\varrho] + \frac{1}{2}\int d^3r d^3r' \frac{\varrho(\mathbf{r})\varrho(\mathbf{r}')}{|\mathbf{r}-\mathbf{r}'|} + E_{XC}[\varrho] , \tag{1.15}$$

where

$$T[\varrho] = (\Psi_o[\varrho], \hat{T}\, \Psi_o[\varrho]) , \tag{1.16}$$

and $\Psi_o[\varrho]$ is the ground-state of an interaction-free electron gas (with $\hat{W}$=0) with an electron density ϱ. (The Hohenberg-Kohn theorem holds independently of the actual form of $\hat{W}$.) In other words, $T[\varrho]$ is the kinetic energy functional of a non-interacting electron gas. Then (1.15) is the definition of the exchange and correation energy functional $E_{XC}[\varrho]$. With these definitions (1.14) takes the form

$$\frac{\delta T}{\delta \varrho} + v_H + v_{XC} = \mu, \tag{1.17}$$

where the Hartree potential v_H is

$$v_H(\mathbf{r}) = v(\mathbf{r}) + \int d^3r' \frac{\varrho(\mathbf{r}')}{|\mathbf{r}-\mathbf{r}'|} , \tag{1.18}$$

and the exchange and correlation potential is defined as

$$v_{XC}(\mathbf{r}) = \frac{\delta E_{XC}}{\delta \varrho(\mathbf{r})} . \tag{1.19}$$

Of course, neither $T[\varrho]$ nor $E_{XC}[\varrho]$ are explicitly known. But on one hand a Thomas-Fermi like approximation for E_{XC} has a safer perspective than for F, and on the other hand a variation of T with respect to a certain wavefunction can be explicitly introduced.

If $\Psi_o[\varrho]$ of (1.16) is a determinant of N_e orthonormalized one-particle wavefunctions $\varphi_i(x_k)$:

$$\Psi_o[\varrho] = \det[\varphi_i(x_k)] , \quad \varrho(\mathbf{r}) = \sum_s \sum_{i=1}^{N_e} |\varphi_i(x)|^2 , \tag{1.20}$$

then (1.16) yields

$$T[\varrho] = \sum_{i=1}^{N_e} (\varphi_i|-\frac{\Delta}{2}|\varphi_i) = T[\varphi_1^*,\dots,\varphi_{N_e}^*;\varphi_1,\dots,\varphi_{N_e}]. \tag{1.21}$$

The second expression of (1.21) may be regarged as an explicitly given functional of the $2N_e$ functions $\varphi_1^*,\dots,\varphi_{N_e}$. If now one of the φ_i is varied with the constraint that the set of the N_e functions φ_i remains orthonormal,

$$(\varphi_i \mid \varphi_j) = \delta_{ij}, \tag{1.22}$$

this leads to a variation

$$\sum_x \frac{\delta T}{\delta\varphi_i^*(x)}\delta\varphi_i^*(x) = (\delta\varphi_i|-\frac{\Delta}{2}|\varphi_i) = \sum_x \frac{\delta T}{\delta\varrho(\mathbf{r})}\varphi_i(x)\delta\varphi_i^*(x)\ . \tag{1.23}$$

In the second equality the variation of the last expression of (1.20)

$$\frac{\delta\varrho(\mathbf{r}')}{\delta\varphi_i^*(x)} = \varphi_i(x)\delta(\mathbf{r}'-\mathbf{r}) \tag{1.24}$$

was used and

$$\sum_x \frac{\delta T}{\delta\varphi_i^*(x)}\delta\varphi_i^*(x) = \sum_x \int d^3r' \frac{\delta T}{\delta\varrho(\mathbf{r}')}\frac{\delta\varrho(\mathbf{r}')}{\delta\varphi_i^*(x)}\delta\varphi_i^*(x). \tag{1.25}$$

Because the variations $\delta\varphi_i^*$ are subject to the constraints (1.22), the second equality (1.23) does not lead to an equation for the functional derivative $\delta T/\delta\varrho$. However, substituting (1.17) into the last expression of (1.23) and adding the constraints (1.22) with Lagrangean multipliers $\varepsilon_{ij}-\mu\delta_{ij}$, so that afterwards the φ_i may independently be varied, we end up with the equations

$$\left[-\frac{\Delta}{2} + v_H + v_{XC}\right]\varphi_i(x) = \sum_j \varepsilon_{ij}\varphi_j(x).$$

As usual, the latter may be simplified to

$$\boxed{\left[-\frac{\Delta}{2} + v_H(\mathbf{r}) + v_{XC}(\mathbf{r})\right]\varphi_i(x) = \varepsilon_i\varphi_i(x)} \tag{1.26}$$

because the orthogonality of the solutions of these equations is automatically obtained. These Kohn-Sham equations are highly non-linear, because both potentials v_H and v_{XC} depend on the φ_i via (1.18)-(1.20).

Though implying an enormous practical progress the step from (1.17) to (1.26) is again pragmatic, since it is known [13] that the determinantal densities (1.20) do not cover all possible densities of non-degenerate ground-states, but it is unknown whether the density minimizing (1.13) is always a determinantal one. Nevertheless, the results so far obtained using (1.26) are by far more then merely encouraging. The presumption that the ground-state be non-degenerate particulary implies that it be not spin-polarized:

$$\varrho(\mathbf{r},+) = \varrho(\mathbf{r},-) = \varrho(\mathbf{r})/2. \tag{1.27}$$

For a short survey of the various spin-dependend and relativistic generalizations of the Hohenberg-Kohn-Sham theory, being equally successful in practical calculations, see [38]. In the present volume we will not tread spin-polarized systems and therefore will not discuss those versions here.

The crucial problem remaining to be solved is now the functional dependence of v_{XC} of (1.19) on the density $\varrho(\mathbf{r})$. As was already mentioned, a Thomas-Fermi like approximation in connection with the advanced state of the theory of the homogeneous electron liquid is used by setting

$$E_{XC}[\varrho] \approx \int d^3r\varrho(\mathbf{r})\varepsilon_{XC}(\varrho(\mathbf{r})), \tag{1.28}$$

where $\varepsilon_{XC}(\varrho)$ is the exchange and correlation energy per particle of a homogeneous electron liquid with constant density ϱ. This local density approximation (LDA) leads via (1.19) to an exchange and correlation potential

$$v_{XC}(\mathbf{r}) = \varepsilon_{XC}(\varrho(\mathbf{r})) + \varrho(\mathbf{r}) \frac{d\varepsilon_{XC}}{d\varrho}(\mathbf{r}) \tag{1.29}$$

locally depending on the ground-state density ϱ. The total energy (1.13) in this LDA reads together with (1.15),(1.21), and (1.28)

$$\boxed{E[v] = \sum_{i=1}^{N_e} (\varphi_i|-\frac{\Delta}{2}|\varphi_i) + \frac{1}{2}\int d^3r(v_H + v + 2\varepsilon_{XC})\varrho,} \tag{1.30}$$

where the φ_i are the N_e lowest energy solutions (understanding spin degeneracy) of (1.26) leading to the density ϱ of (1.20). With the help of the Kohn-Sham equations (1.26), this may be rewritten in various ways, e.g.,

$$\boxed{E[v] = \sum_{i=1}^{N_e} \varepsilon_i + \frac{1}{2}\int d^3r(v - v_H - 2v_{XC} + 2\varepsilon_{XC})\varrho.} \tag{1.31}$$

For extended systems, the term $(1/2)\int d^3rv\varrho$ must be combined with the first term of (1.8) in order to give a finite result per unit volume.

Neglecting correlation effects in the homogeneous electron liquid, that is using a single determinant of plane waves instead of the ground-state, the exchange-only expressions

$$\varepsilon_X(\varrho) = -\frac{3}{4}\left(\frac{3\varrho}{\pi}\right)^{1/3} \tag{1.32}$$

and

$$v_X(\mathbf{r}) = -\left(\frac{3\varrho(\mathbf{r})}{\pi}\right)^{1/3} \tag{1.33}$$

are easily obtained. At metallic densities, (1.33) is the main contribution to (1.29), the latter is therefore usually written as

$$v_{XC}(\mathbf{r}) = (3/2)\alpha(\varrho(\mathbf{r}))v_X(\mathbf{r}), \tag{1.34}$$

which is a definition of a factor $\alpha(\varrho)$[1]. Many calculations have been performed with a constant factor α varying only with the atomic species [110,111]. The most popular and successful version of $\alpha(\varrho)$ is however that of Hedin and Lundqvist [59] obtained by interpolating numerical results of (1.29) where ε_{XC} was calculated from one of the best results for the dielectric function of the homogenous electron liquid. It may be written as

$$\alpha_{HL}(r_s) = 2/3 + 0.0246r_s\ln(1 + 21/r_s) \tag{1.35}$$

where ϱ is replaced by the density parameter r_s defined as

$$\frac{4\pi}{3}r_s^3 = 1/\varrho\ . \tag{1.36}$$

1) The extra factor (3/2) in (1.34) has a mere historical origin and is only preserved here in order to facilitate comparison with the original literature.

The value of α from (1.35) varies between 2/3 and 1.184, monotonically increasing with r_s. It corresponds to an exchange and correlation energy per particle

$$\varepsilon_{XC}^{HL}(r_s) = -\frac{3}{4}\left(\frac{3}{2\pi}\right)^{2/3}\frac{1}{r_s} - 0.0225\left[\left(1+\left(\frac{r_s}{21}\right)^3\right)\ln\left(1+\frac{21}{r_s}\right)-\left(\frac{r_s}{21}\right)^2+\frac{r_s}{42}-\frac{1}{3}\right] \tag{1.37}$$

to be used in (1.30) or (1.31).

Using this scheme, the potential functions (1.8) have been calculated successfully for a great number of molecules (see e.g. [67,98]). For crystalline solids, where the vectors $\mathbf{R}_a$ of (1.8) form a regular lattice, U(a) may be calculated as a function of the lattice constant a yielding equilibrium lattice constants, binding energies and elastic moduli. For instance the equilibrium lattice constants are generally obtained within a few percent deviation from experimental values. For a systematic study of metals see [96]. In the case of semiconductors practically the full dependence $U(\mathbf{R})$ has been computed yielding the lattice dynamics and even quantitatively lattice anharmonicities [24,79,80].

It is important to note that in all these application total energy differences of the same system in different atomic arrangements are calculated. There is no doubt that the absolute value of the total energy comes out incorrectly as there is little justification for the LDA in the core region of the atoms where the electron density is extremely inhomogeneous. These errors, however, cancel in taking energy differences as the atomic cores do not take part in the chemical binding. For attempts going beyond LDA see [51].

Using the total energy expressions (1.30) or (1.31) it is also possible to calculate energy differences corresponding to changes in particle number. If we denote the ground-state energy of a system of N_e electrons in the external potential v by $E_{N_e}[v]$, then the electron affinity A is given by

$$-A = E_{N_e+1}[v] - E_{N_e}[v], \tag{1.38}$$

and the first ionization energy is

$$I = E_{N_e-1}[v] - E_{N_e}[v]. \tag{1.39}$$

Denoting by ε_i, $\varrho, \ldots$ the Kohn-Sham energies, particle density, and so on of the N_e-electron system and by $\delta\varepsilon_i$, $\delta\varrho, \ldots$ their differences between the (N_e+1)-electron and the N_e-electron systems, we obtain from (1.31)

$$E_{N_e+1}[v] - E_{N_e}[v] = \varepsilon_{N_e+1} + \sum_{i=1}^{N_e} \delta\varepsilon_i +$$

$$+ \frac{1}{2}\,\delta\int d^3r(v - v_H - 2v_{XC} + 2\varepsilon_{XC})\varrho.$$

For large systems, this expression may be estimated to first order in $\delta\varepsilon_i$, $\delta\varrho, \ldots$. Equation (1.26) yields in the first perturbation order

$$\sum_{i=1}^{N_e} \delta\varepsilon_i = \sum_{i=1}^{N_e} (\delta v_H + \delta v_{XC})|\varphi_i|^2 = (\delta v_H + \delta v_{XC})\varrho,$$

where (1.20) was used in the last equation. Furthermore, from (1.18),

$$\varrho\delta v_H = (v_H - v)\delta\varrho,$$

and from (1.19) and (1.28)

$$\int d^3r v_{XC}\,\delta\varrho = \delta E_{XC} = \delta\int d^3r\varepsilon_{XC}\varrho.$$

Combining these relations we find

$$\boxed{E_{N_e+1}[v] - E_{N_e}[v] = \varepsilon_{N_e+1} + N_e O([\delta\varrho/\varrho]^2),} \tag{1.40}$$

where $O(x)$ means the order of x, and

$$\delta\varrho/\varrho \approx |\varphi_{N_e+1}|^2 \Big/ \sum_{i=1}^{N_e} |\varphi_i|^2. \tag{1.41}$$

For an N-atom system, the correction term of (1.40) is proportional to N^{-1}, hence it vanishes for an extended system.

In the case of a metal the electron affinity and the first ionization energy are equal to each other and equal to the work function W:

$$-\varepsilon_{N_e+1} = A = I = -\varepsilon_{N_e} = -\varepsilon_F = W. \tag{1.42}$$

These relations presuppose, of course, that the Hartree potential v_H in (1.26) includes the surface dipole barrier at the metal surface and is normalized to be zero at infinity outside of the metal. If this last presumption is not observed as usual in band calculations, the more general relation

$$W = D - \varepsilon_F \,, \qquad D = v_H(\infty) \tag{1.43}$$

must be used, where the surface dipole barrier D is the difference between the zero of v_H and its value at infinity outside of the metal. Note, that (1.42) also implies that the Kohn-Sham equations do not yield an energy gap at the Fermi level of a metal.

In the case of a semiconductor or insulator the fundamental energy gap may be defined as

$$\Delta = I - A. \tag{1.44}$$

From (1.38)-(1.40) one would obtain

$$\Delta = \Delta_{KS} = \varepsilon_{N_e+1} - \varepsilon_{N_e}, \tag{1.45}$$

that is, the fundamental Kohn-Sham gap should be equal to the experimentally observed one. Unfortunately this is not found in actual calculations, and there are strong indications that this defect is not caused by the inaccurate knowledge of $E_{XC}[\varrho]$ and the use of the LDA. The gap itself seems to produce a mathematical complication. By definition, the two energies on the left hand side of (1.40) are the true ground-state energies corresponding to the densities $\varrho+\delta\varrho$ and ϱ, respectively, where $\delta\varrho$ and ϱ are given by the numerator and denominator of (1.41) provided the Kohn-Sham equations are correct. Hence, for an extended system, (1.40) may be rewritten as

$$\int d^3r \, \frac{\delta E[\varrho]}{\delta\varrho} \, |\varphi_{N_e+1}|^2 = \varepsilon_{N_e+1}, \quad \varrho = \sum_{i=1}^{N_e} |\varphi_i|^2. \tag{1.46}$$

Due to (1.12) this is eqivalent to

$$\int d^3r \left[\frac{\delta F[\varrho]}{\delta\varrho} + v\right] |\varphi_{N_e+1}|^2 = \varepsilon_{N_e+1}. \tag{1.47}$$

On the other hand, for any ground-state density, (1.14) should be fulfilled implying

$$\int d^3r \left[\frac{\delta F[\varrho]}{\delta\varrho} + v\right] |\varphi_{N_e+1}|^2 = \mu \tag{1.48}$$

since φ_{N_e+1} is normalized to unity. This is a contradiction, if an energy gap is present at the Fermi level, because in this case the right hand sides of (1.47) and (1.48) are different from each other. Since the only open assumption used in these considerations is the existence of the functional derivative $\delta F[\varrho]/\delta\varrho$, the conclusion should be that this derivative does not exist in the considered case [102,112,36].

The situation is, however, sufficiently subtle, especially in the light of the very encouraging results for total energy variations with respect to atomic displacements of the N_e-electron system [24,79,80], as mentioned earlier. Though the problem is not yet finally resosved, the correct Kohn-Sham results for $U(\mathbf{R})$ do not necessarily contradict to the failure in the case of the gap [34].

1.3. Quasi-particles

Next to the electronic ground-state properties as discussed in the preceding section the electronic excitations are of great interest determining largely the thermodynamic and kinetic properties of the system. (The vibrational ionic excitations are, due to the adiabatic theorem, basically governed by the electronic ground-state properties as was already discussed.) In the case of a finite system - an atom or a molecule - the stationary excitations, experimentally seen in the line spectra, are to be considered. The density of electronic stationary states on the energy scale is roughly

$$D(E) = \frac{1}{\Delta E} \approx 2^{N_e} \text{ states per Rydberg,} \tag{1.49}$$

where ΔE is the energetic distance of two successive stationary states. For extended systems $N_e \sim 10^{23}$, and $\Delta E \approx 10^{-10^{23}}$ Ryd becomes so exorbitantly small, that there is not the least hope to experimentally single out one definite stationary state. The stationary states lose their experimental and theoretical importance. What is seen in experiment, and, hence, what quantum theory of macrosystems should be concerned with, are not the stationary but the quasi-stationary (long-lived but damped) states being excited by bombarding the system with various test particles in experiment and being created by properly chosen field operators in theory.

Consider an electronic macrosystem (with the Hamiltonian (1.2)) in a ground-state Ψ_o and inject at the time t=0 an additional electron:

$$t = 0: \quad |\Psi>_{t=0} = \hat{c}^+_\alpha(0)|\Psi_o>, \tag{1.50}$$

where $\hat{c}^+_\alpha(t)$ means the electron creation operator in Heisenberg representation depending on parameters α (e.g. momentum or position and spin polarization), and $|\Psi>$ means the Dirac state vector [121]. If $|\Psi>$ were a stationary state with as excitation energy ξ_α, than, at t>0, the system would be found in the state

$$t > 0: \quad |\Psi>_t = \hat{c}^+_\alpha(t)|\Psi_o> = e^{-i\xi_\alpha t}|\Psi>_{t=0}. \tag{1.51}$$

If Ψ is only quasi-stationary, it decays in time, and the probability

$$w\,(t) = |<\Psi_o|\hat{c}_\alpha(t)\hat{c}^+_\alpha(0)|\Psi_o>|^2 \tag{1.52}$$

for the system to be in the state (1.50) at t>0 is smaller then one. If Ψ_o was an N_e-particle state of the system, then Ψ is an (N_e+1)-particle state. As is well known, an exited state of the system with N_e-1 particles can also be obtained from Ψ_o by acting on it with an annihilation operator $\hat{c}_\alpha$ leaving a "hole" in the one-particle state α. In order to treat both cases simultaneously, the latter action is supposed to have taken place at some t<0 and its result is observed at t=0, thus obtaining

$$w_\alpha(|t|) = |G_\alpha(t)|^2,$$

$$G_\alpha(t) = -i<\Psi_o|\ T\ \hat{c}_\alpha(t)\hat{c}^+_\alpha(0)|\Psi_o> =$$

$$= \begin{cases} -i<\Psi_o|\hat{c}_\alpha(t)\hat{c}^+_\alpha(0)|\Psi_o> & \text{for } t > 0 \\ i<\Psi_o|\hat{c}^+_\alpha(0)\hat{c}_\alpha(t)|\Psi_o> & \text{for } t < 0. \end{cases} \tag{1.53}$$

Here, T is Wieck's time ordering operator defined by the last equation, and $G_\alpha(t)$ is the electron Green's function. Due to the fact that we have a situation with variable particle number, time dependencies are determined by the Hamiltonian

$$\hat{\Omega} = \hat{H} - \mu\hat{N}, \tag{1.54}$$

where μ is the chemical potential and $\hat{N}$ the particle number operator.

In a system of non-interacting particles (with Hamiltonian $\hat{H}_o$) we would have $\xi_\alpha=\varepsilon_\alpha-\mu$ in (1.51), where ε_α is the energy of the created one-particle stationary state corresponding to the Hamiltonian $\hat{H}_o$. Hence, in this case,

$$G_\alpha(t) = \begin{cases} -ie^{+i(\varepsilon_\alpha-\mu)t} & \text{for} \quad t > 0,\ \varepsilon_\alpha > \mu \\ +ie^{-i(\varepsilon_\alpha-\mu)t} & \phantom{\text{for}} \quad t < 0,\ \varepsilon_\alpha < \mu \\ 0 & \phantom{\text{for}} \quad \text{elsewhere.} \end{cases} \tag{1.55}$$

A Fourier transformation

$$G_\alpha(\omega) = \int_{-\infty}^{\infty} dt e^{i\omega t} G_\alpha(t) \tag{1.56}$$

of the expression (1.55) yields

$$G_\alpha(\omega \pm i0) = \frac{1}{\omega - \varepsilon_\alpha + \mu \pm i0} \quad \text{for} \quad \varepsilon_\alpha \gtrless \mu. \tag{1.57}$$

If, at t=0, $\hat{c}^+_\alpha$ does not create a stationary state Ψ but only a quasi-stationary one, then, at t>0, $|\Psi>_t$ is no longer given by the last expression (1.51). Instead, this contribution to $|\Psi>_t$ will decay in time according to the law

$$e^{-t/\tau_\alpha} \tag{1.58}$$

with the lifetime τ_α of the quasi-stationary state. Consequently, the Green's function (1.53) also contains (among others) terms (1.55) with certain coefficients, but with complex energies ε_α, where

$$\text{Im}\ \varepsilon_\alpha = -\text{sign}(\text{Re}\ \varepsilon_\alpha - \mu)/\tau_\alpha. \tag{1.59}$$

This means that $G_\alpha(\omega)$ contains a term[1)]

$$G_\alpha(\omega) = \frac{Z_\alpha}{\omega - \varepsilon_\alpha + \mu} + \ldots \tag{1.60}$$

---

1) Our mere phenomenological considerations do not indicate the occurence of a factor Z_α different from unity in (1.60). The point is that the complete quasi-particle creation operator is

$$\hat{c}^{(q)+}_\alpha = \sqrt{Z_\alpha}\ \hat{c}^+_\alpha + \text{something},$$

where "something" creates the polarization cloud surrounding an "electron quasi-particle" in a solid. $Z_\alpha<1$ is the so-called renormalization constant for the field amplitude.

and therefore

$$\boxed{G_{\alpha}^{-1}(\varepsilon_{\alpha} - \mu) = 0.} \tag{1.61}$$

This equation together with the definitions (1.53) and (1.56) gives the energy of a quasi-stationary electron excitation as a function of the parameters α [9,86].

Strictly speaking, there is no proof for the decay of the quasi-stationary excitations to follow a mere exponential law (1.58), but it can be expected that in a sufficiently long interval of time $\Delta t \gtrsim \tau_{\alpha}$ (1.58) describes the decay in a sufficiently precise manner. As a consequence, the Green's function $G_{\alpha}(\omega)$ has a more complicated - in general non-meromorphic - structure compared to (1.60), but on the real axis of the complex ω-plane it may be well approximated by the expression (1.60). It should be noted in this connection that the whole quasi-particle concept, based on the analytical structure of (1.60), rests on this realistic assumption suggested by experimental facts with the help of theoretical reasoning along the lines indicated. Of course, for a quasi-stationary excitation,

$$|\mathrm{Im}\,\varepsilon_{\alpha}| \ll |\mathrm{Re}\,\varepsilon_{\alpha} - \mu| \tag{1.62}$$

is supposed.

In the important special case when the set α of parameters contains the momentum of the excitation of a translationally invariant system or the quasi-momentum $\mathbf{k}$ mod $\mathbf{G}$ in a crystalline lattice with $\mathbf{G}$ as a vector of the reciprocal lattice, that is, (ν denotes the remaining parameters as band and spin indices)

$$\varepsilon_{\alpha} \equiv \varepsilon_{\nu}(\mathbf{k}), \tag{1.63}$$

one speaks of quasi-particles and their dispersion relation (1.63). In this terminology the state Ψ_0 of (1.53) may be characterized as the state without quasi-particles or the quasi-particle vacuum. As the temperature is gradually increased, quasi-particles are created, but their number is in the beginning rather small so that the solid may be treated as a dilute gas of weakly interacting quasi-particles with an arbitrary dispersion law.[1)]

---

1) The quasi-particle concept for the description of the excitation spectrum of condensed matter was first introduced by L.D. Landau in (cont. on p. 24)

As long as the number of quasi-particles is sufficiently small and their mutual interaction is weak, the state of the system may be well characterized by the mean occupation numbers n_α of the quasi-particle states. The total energy of the system may then be expressed as a functional $E(n_\alpha)$ of these numbers, and

$$\mathrm{Re}\,\varepsilon_\alpha = \frac{\delta E}{\delta n_\alpha} . \tag{1.64}$$

The quasi-particle interaction leads in this approach to a non-linear dependence of the energy functional on the n_α and hence to a dependence of the quasi-particle energy $\mathrm{Re}\,\varepsilon_\alpha$ on the averaged numbers of quasi-particles present, that is, on the state of the system:

$$\frac{\delta(\mathrm{Re}\,\varepsilon_\alpha)}{\delta n_\beta} = \frac{\delta^2 E}{\delta n_\beta \delta n_\alpha}, \quad \delta(\mathrm{Re}\,\varepsilon_\alpha) = \int d\beta \delta n_\beta \frac{\delta^2 E}{\delta n_\beta \delta n_\alpha} . \tag{1.65}$$

Just on the formulas (1.64) and (1.65) is based the phenomenological Landau theory.

We consider now again a system of non-interacting electrons in the one-particle Hartree potential $v_H(\mathbf{r})$ of (1.18) and introduce its electron Green's function in real space representation by

$$G^{(o)}(t,x,x') = -i\langle \Psi_o | \, T \hat{\psi}(t,x) \, \hat{\psi}^+(0,x') | \Psi_o \rangle \tag{1.66}$$

(it must be chosen non-diagonal in x since, contrary to α of (1.53), $x=(\mathbf{r},s)$ is "not a good quantum number"). From the equation of motion of the field operators $\hat{\psi}(t,x)$ it is easy to find

$$\left[\omega + \mu + \frac{\Delta}{2} - v_H(\mathbf{r})\right] G^{(o)}(\omega,x,x') = \delta(x-x'). \tag{1.67}$$

(cont. from p. 23)
1941 [82] and successfully applied to explain superfluidity and the thermodynamic properties of the superfluid liquid He^4. Later on, in 1956 [83], he applied this concept to describe the properties of the normal (i.e. not superfluid) Fermi liquid He^3. In 1958 B.M. Galitzky and A.B. Migdal [44] derived the consistent microscopic theory of quasi-particles using the apparatus of Green's functions. The quasi-particle concept has proved to be fundamental for the whole condensed matter theory [86].

In an operator notation with respect to the x-dependence this may be written as

$$\hat{G}^{(o)-1}(\omega) = \omega + \mu + \frac{\nabla^2}{2} - v_H(\mathbf{r}). \tag{1.68}$$

For the system of interacting electrons in the corresponding external potential $v(\mathbf{r})$ the so-called self-energy operator $\hat{\Sigma}(\omega)$ is indroduced by

$$\hat{\Sigma}(\omega) = \hat{G}^{-1}(\omega) - \hat{G}^{(o)-1}(\omega) \tag{1.69}$$

or

$$\left[\hat{G}^{(o)-1}(\omega) + \hat{\Sigma}(\omega)\right]\hat{G}(\omega) = 1. \tag{1.70}$$

In full notation this means ($\int dx \equiv \sum_{s=-1/2}^{1/2} \int d^3r$)

$$\left[\omega + \mu + \frac{\Delta}{2} - v_H(\mathbf{r})\right] G(\omega, x, x') - \\ - \int dx'' \, \Sigma(\omega, x, x'') G(\omega, x'', x') = \delta(x-x') \tag{1.71}$$

In the periodic potential of a crystalline lattice the quasi-momentum of the electron is conserved, and therefore the Green's function is diagonal in the quasi-momentum $\mathbf{k}$ in the corresponding representatión:

$$\delta(\mathbf{k}-\mathbf{k}')G_{\mathbf{GG}'}(\omega,\mathbf{k}) = \int dx dx' \, e^{i(\mathbf{k}+\mathbf{G})\mathbf{r}} \, G(\omega, x, x') \, e^{-i(\mathbf{k}'+\mathbf{G}')\mathbf{r}'}. \tag{1.72}$$

In the vicinity of the solution of (1.61) (in the complex ω-plane) G takes the form (1.60):

$$G_{\mathbf{GG}'}(\omega,\mathbf{k}) = \sum_{\nu} \frac{Z^{(\nu)}_{\mathbf{GG}'}(\mathbf{k})}{\omega + \mu - \varepsilon_\nu(\mathbf{k})} + g_{\mathbf{GG}'}(\omega,\mathbf{k}), \tag{1.73}$$

where $g_{\mathbf{GG}'}$ has already no singularity for $\omega \to \varepsilon_\nu(\mathbf{k})-\mu$. Substituting (1.73) together with (1.72) into the equation (1.71) and taking the limit $\omega \to \varepsilon_\nu(\mathbf{k})-\mu$, one obtains

$$\left[\varepsilon_\nu(\mathbf{k}) + \frac{\Delta}{2} - v_H(\mathbf{r})\right] Z^{(\nu)}_{\mathbf{k}}(x, x') - \\ - \int dx'' \, \Sigma(\varepsilon_\nu(\mathbf{k})-\mu, x, x'') \, Z^{(\nu)}_{\mathbf{k}}(x'', x') = 0 \tag{1.74}$$

Here, Z carries the additional index $\mathbf{k}$ since this function is the Fourier transform of the residuum of (1.73) connected with the pole at $\varepsilon_\nu(\mathbf{k})$. The variable x' enters the linear equation (1.74) only as a parameter, hence one may put

$$Z_{\mathbf{k}}^{(\nu)}(x,x') \equiv \varphi_{\mathbf{k}\nu}(x)\chi_{\mathbf{k}\nu}(x'). \tag{1.75}$$

After substituting this into (1.74) the "Schrödinger equation"

$$\boxed{\left[-\frac{\Delta}{2} + v_H(\mathbf{r})\right]\varphi_{\mathbf{k}\nu}(x) + \int dx'\,\Sigma(\varepsilon_\nu(\mathbf{k})-\mu,x,x')\,\varphi_{\mathbf{k}\nu}(x') = \varepsilon_\nu(\mathbf{k})\varphi_{\mathbf{k}\nu}(x)} \tag{1.76}$$

is obtained. Introducing at $\varepsilon_\nu(\mathbf{k}) \approx \mu$ the local approximation [59]

$$\int dx' \operatorname{Re}\Sigma(\varepsilon_\nu(\mathbf{k}) - \mu,x,x')\varphi_{\mathbf{k}\nu}(x') \approx v_\Sigma(x)\varphi_{\mathbf{k}\nu}(x) \tag{1.77}$$

for the self-energy operator and neglecting its imaginary part (which goes to zero for $\varepsilon_\nu(\mathbf{k}) \to \mu$) one ends up with the usual Schrödinger equation

$$\boxed{\left[-\frac{\Delta}{2} + v_H(\mathbf{r}) + v_\Sigma(x)\right]\varphi_{\mathbf{k}\nu}(x) = \varepsilon_\nu(\mathbf{k})\varphi_{\mathbf{k}\nu}(x).} \tag{1.78}$$

The solution of (1.76) or its approximate variant (1.78) should give the bandstructure of quasi-particle excitations, where for "particle excitations" with $\operatorname{Re}\varepsilon_\nu(\mathbf{k}) > \mu$ the excitation energy is $\xi_\nu(\mathbf{k}) = \operatorname{Re}\varepsilon_\nu(\mathbf{k}) - \mu$ and for "hole excitations" with $\operatorname{Re}\varepsilon_\nu(\mathbf{k}) < \mu$ the excitation energy is $\xi_\nu(\mathbf{k}) = \mu - \operatorname{Re}\varepsilon_\nu(\mathbf{k})$[1]. Keeping in mind the conservation of the total quasi-momentum in electron-electron (or electron-phonon) collisions it is easy to show [86] that, for T=0 and at low excitation energies, due to the electron-electron interaction

$$\operatorname{Im}\varepsilon_\nu(\mathbf{k}) \sim (\operatorname{Re}\varepsilon_\nu(\mathbf{k}) - \mu)^2 \tag{1.79}$$

---

[1] Note that a particle may be added to or removed from a macroscopic system without changing the thermodynamic state, if the energy of the particle is equal to the chemical potential μ. The "particle excitation" may be thought of being obtained by first adding an electron with energy μ without any excitation and then exciting it to the energy $\operatorname{Re}\varepsilon_\nu(\mathbf{k}) > \mu$. Analogonsly a "hole excitation may be obtained by first exciting an electron from $\operatorname{Re}\varepsilon_\nu(\mathbf{k})$ to the chemical potential μ and then removing it.

and due to the electron-phonon interaction

$$\operatorname{Im} \varepsilon_\nu(\mathbf{k}) \sim |\operatorname{Re} \varepsilon_\nu(\mathbf{k}) - \mu|^3 \tag{1.80}$$

holds.

In the model of non-interacting with each other particles, the Fermi surface

$$\varepsilon_\nu(\mathbf{k}_F) = \varepsilon_F = \mu(T=0) \tag{1.81}$$

of a metal is connected with the total particle number N_e in a very simple way: there must be just N_e one-particle levels (understanding the multiplicity of possible degeneracies including spin degeneracy) below the Fermi level. If τ_F is the volume in $\mathbf{k}$-space "inside" of the Fermi surface, that is the volume of the region where $\operatorname{Re} \varepsilon_\nu(\mathbf{k}) < \varepsilon_F$ and summation over all branches ν of the dispersion law (including the spin index) to be understood, then

$$\frac{N_e}{\vartheta} = \frac{\tau_F}{(2\pi)^3} \tag{1.82}$$

with the volume ϑ of the metal. For real systems of interacting particles such a connection is not evident by itself. However, Landau's hypothesis, that, for the fermion excitation spectra, (1.82) holds also in this case, where now N_e is the number of original particles and τ_F the volume of hole-excitation quasi-momenta, was rigorously demonstrated by Luttinger [93].

The relation (1.82) uniquely determines the Fermi surface (1.81) of the quasi-particle spectrum $\varepsilon_\nu(\mathbf{k})$ obtained from (1.76) (or, with use of the local approximation (1.77), from (1.78)) and seen in experiment. According to (1.42), the same relation (1.82) also determines the "Fermi surface of the Kohn-Sham spectrum $\varepsilon_\nu(\mathbf{k})$" obtained from (1.26) (for a periodic potential v_H+v_{XC}). The present author is, however, not aware of any rigorous proof that both Fermi surfaces should coincide (cf. also the discussion of this point in [59]). In the local approximation (1.77) v depends on x only via the density $\varrho(x)$. The function $v_\Sigma(\varepsilon-\mu;\varrho)$ may be written in analogy to (1.34) as

$$v_\Sigma(\varepsilon - \mu; \varrho) = (3/2)\alpha_\Sigma(\varepsilon - \mu; r_s) v_x(\mathbf{r}). \tag{1.83}$$

Numerical values for v_Σ (as a function of k instead of ε; ε and k being uniquely linked to each other by the quasi-particle dispersion $\varepsilon(k,\varrho)$ in the homogeneous electron liquid of density ϱ) are given in [19]. From these values, $\alpha_{\Sigma}(\varepsilon-\mu;r_s)$ may be extracted and compared to $\alpha_{HL}(r_s)$ of (1.35). From fig. 1 one can see that $\alpha_{\Sigma}(0;r_s)$ and $\alpha_{HL}(r_s)$ are very close to each other in the whole range of metallic densities. This explains why Fermi surfaces from density functional calculations are so close to experimental ones.

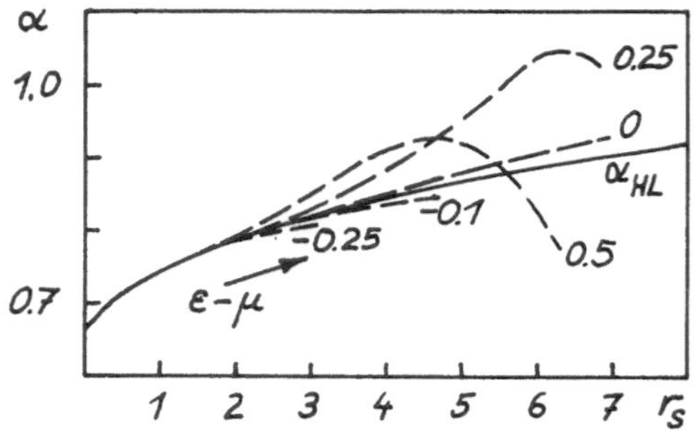

Fig. 1. Comparison of $\alpha_{\Sigma}(\varepsilon-\mu;r_s)$ of (1.83) as obtained from the tables of v_{Σ} given in [59] for $\varepsilon-\mu= -0.25, -0.1, 0, 0.25, 0.5$ Hartree with $\alpha_{HL}(r_s)$ of (1.35). The values of $\alpha_{\Sigma}(0;r_s)$ and $\alpha_{HL}(r_s)$ are very close to each other in the whole range of metallic densities.

Solving the equation (1.76) with the so-called G-W-approximation [58] for the self-energy operator $\hat{\Sigma}$, thereat using the ground-state electron density as obtained from a self-consistent solution of (1.26), Hybertson and Louie [62] obtained for semiconductors and insulators a band-structure $\varepsilon_\nu(\mathbf{k})$ in excellent agreement with experiment and without any discrepancy between the calculated and the experimental fundamental gap.

2. THE LCAO FORMALISM

In this and the following chapters we are concerned with the numerical solution of the one-particle equations (1.26) and (1.76) or (1.78), respectively. The "one-particle potential" will be denoted by

$$v = v_H(\mathbf{r}) + v_{XC}(\mathbf{r}) \tag{2.1}$$

and

$$v = v_H(\mathbf{r}) + \hat{\Sigma} \quad \text{or} \quad v = v_H(\mathbf{r}) + v_{\Sigma}(x), \tag{2.2}$$

respectively. Both the qualitative discussions and the numerical methods applied in what follows are very little affected by whether the potential v is local or non-local (like the integral operator $\hat{\Sigma}$) and whether it depends on the position $\mathbf{r}$ of the electron only or on position and spin polarization $x=(\mathbf{r},s)$. It is one advantage of variational methods like LCAO that they apply in an equal manner to local and non-local potentials. Keeping this in mind, we restrict, without loss of generality, the whole treatment in the following to the simplest case of a local potential $v(\mathbf{r})$.

Hence, we consider the problem

$$\hat{h}\varphi_\nu(\mathbf{r}) = \varepsilon_\nu\varphi_\nu(\mathbf{r}), \qquad \hat{h} = -\frac{\Delta}{2} + v(\mathbf{r}). \tag{2.3}$$

The method applied makes explicit use of the shape of $v(\mathbf{r})$ as a certain geometrical arrangement of atomic potential wells. It is the most natural idea then to construct the wavefunction $\varphi_\nu(\mathbf{r})$ as a superposition of atom-like wavefunctions in the same geometrical arrangement. That is basically all what characterizes the LCAO method.

We start the discussion of this approach by recalling some important features of atomic wavefunctions and providing the numerics for computing them. Then, in the second section the basic LCAO equations are derived. Emphasis is put on an economic treatment of the orthogonalization to core states. In a subsequent section the Bloch state representation for regular crystalline lattices is introduced and the account for an inversional symmetry is outlined in detail. Finally, in the last section the computation of the appearing multi-centre integrals is described for the case of a Slater-type basis. Thus, the reader will be provided with all important details for performing an LCAO calculation for a given potential and a given basis of atom-like states, both expressed through Slater-type functions, as well as with references of alternative approaches. Subsequent chapters deal with how to find an optimum atom-like basis and how to obtain a self-consistent crystal potential.

2.1. Atomic one-particle wavefunctions

Consider the problem

$$\hat{h}_{at}\varphi_i^{at} = \varepsilon_i^{at}\varphi_i^{at}, \qquad \hat{h}_{at} = -\frac{\Delta}{2} + v_{at}(r) \tag{2.4}$$

with a spherically symmetric atomic potential

$$-\frac{A}{r} \xleftarrow[r\to 0]{} v_{at}(r) \xrightarrow[r>r^{at}]{} -\frac{1}{r}, \qquad r^{at} \approx 2\ldots 3 \text{ a.u.} \tag{2.5}$$

Near the nucleus, that is for $r \to 0$, the potential is dominated by the Coulomb potential of the electric charge A of the nucleus, and outside of the atom radius r^{at}, typically being as large as 2 to 3 Bohr radii, it becomes equal to the Coulomb potential of a single charged ion because an electron, moving out of a neutral atom, leaves behind itself a positive single charged ion.

Due to the spherical symmetry of the potential, the problem may be separated [85] in spherical coordinates. Then, the wavefunction takes the form

$$\varphi_i^{at}(\mathbf{r}) = \varphi_{nl}(r)Y_{lm}(\hat{\mathbf{r}}) \tag{2.6}$$

with the spherical harmonic Y_{lm} corresponding to the angular momentum quantum number l and the azimut quantum number m (see Appendix 1.). For the radial wavefunction $\varphi_{nl}(r)$ the equation

$$\left[-\frac{1}{r}\frac{d^2}{dr^2}r + \frac{l(l+1)}{r^2} + 2v^{at}(r)\right]\varphi_{nl}(r) = 2\varepsilon_{nl}^{at}\varphi_{nl}(r) \tag{2.7}$$

is obtained. Multiplying this equation by r one finds

$$\frac{d^2u_{nl}}{dr^2} = \left[\frac{l(l+1)}{r^2} + 2v^{at}(r) - 2\varepsilon_{nl}^{at}\right]u_{nl}, \quad u_{nl} \equiv r\varphi_{nl}. \tag{2.8}$$

The expression in brackets of (2.8) is the relative curvature of the function $u_{nl}(r)$. The normalizable solutions we are interested in are those with

$$0 \underset{r\to 0}{\longleftarrow} u_{nl}(r) \underset{r\to\infty}{\longrightarrow} 0. \tag{2.9}$$

For a given angular momentum l, with increasing energy eigenvalues ε^{at}_{nl} they are in the well known manner characterized by a successively increasing number of nodes (if the node of u_{nl} at r=0 is counted accordant to its multiplicity, then this function has exactly n nodes). Since the first two terms in brackets of (2.8) diverge for $r\to 0$, ε^{at}_{nl} may be neglected against these terms for sufficiently small r. For atomic problems this is generally the case for r-values inside of the outermost node of the radial wavefunction. Consequently, for r-values inside of the outermost node of $\varphi_{nl}(r)$, all wavefunctions $\varphi_{n'l}(r)$ with n'>n are practically proportional to φ_{nl}. This is a very important property of atomic wavefunctions shown in fig. 2.1. Especially the inner nodes do practically not move, when n increases. Physically this is due to the fact that the kinetic energy of an electron moving in this region exceeds $|\varepsilon^{at}_{n'l}|$ typically by several orders of magnitude.

This behaviour has important consequences, if atomic eigenfunctions $\varphi^{at}_{i}(\mathbf{r})$ are used as basis functions for the expansion of other wavefunctions. Let, e.g., the first m functions φ_{nl}, n=1,...,m be used as a finite basis to expand another radial function ψ_l with the angular quantum number l. Let the outermost node of φ_{ml} be at r_m and let ψ_l be a bound eigenstate of (2.7) with a potential v(r), different from $v^{at}(r)$ only outside of r_m. Then the expansion of ψ_l in terms of the φ_{nl} practically exactly reproduces ψ_l inside of r_m. On the other hand, since the atomic eigenvalues ε^{at}_{nl} accumulate at ε=0, no basis of functions φ_{nl} will be sufficiently complete to reproduce ψ_l as a solution of (2.7) with a positive energy in a range, where $|v^{at}(r)|$ is of the same order as or smaller then the energy of the state ψ_l. This is so independently on how many basis functions φ_{nl} are used.

In order to characterize the atomic wavefunctions more quantitatively consider the expression

$$b_{l\alpha} = r^{l+1}\, e^{-\alpha r},$$

for which

$$\frac{d^2 b_{l\alpha}}{dr^2} = \left[\frac{l(l+1)}{r^2} - \frac{2\alpha(l+1)}{r} + \alpha^2\right] b_{l\alpha}$$

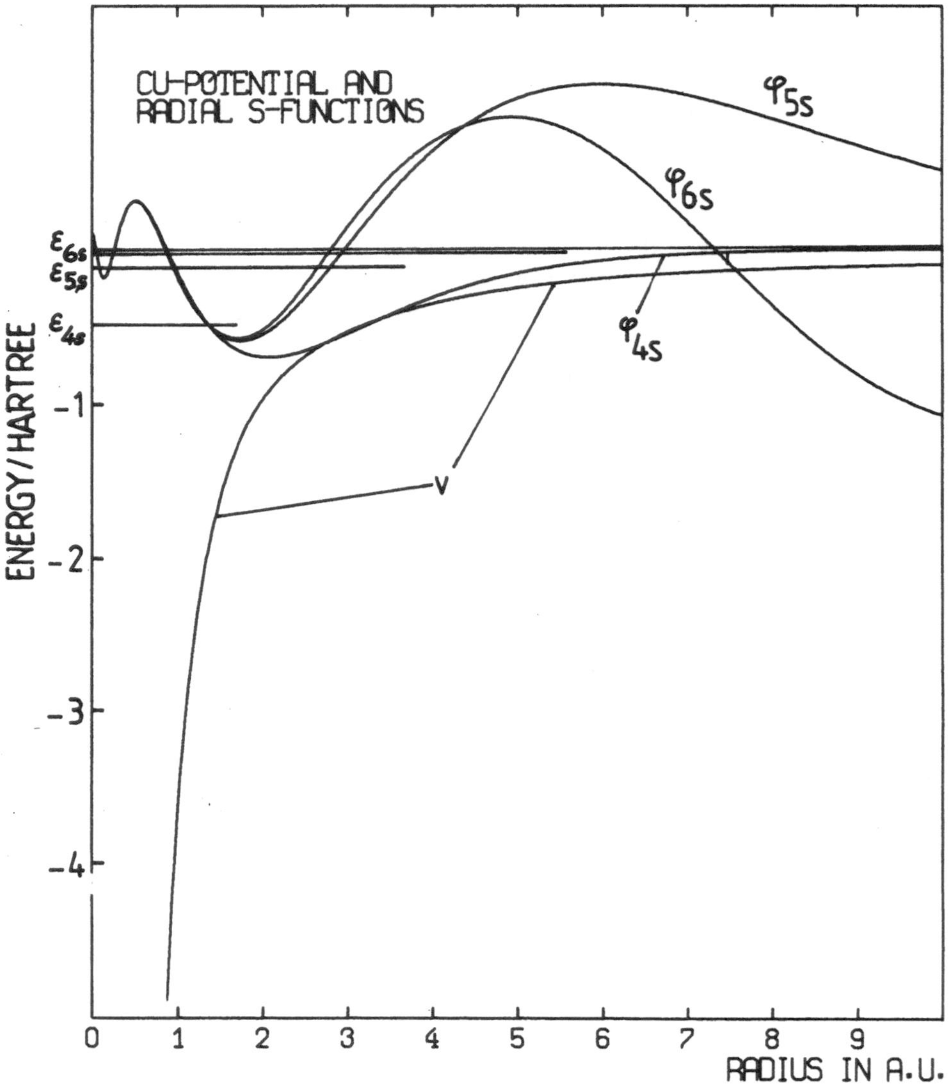

Fig. 2.1. Ionic potential and atomic radial s-functions $u_{ns}(r)$ of the Cu atom. The wavefunctions are normalized to coincide for small r. The corresponding energy levels ε_{ns} are also indicated. For r-values inside of the outermost node of the 4s-function (r<0.87) the three plotted wavefunctions practically coincide.

holds. Comparing this relation with (2.8) one finds, that for large r the function $b_{l\alpha}$ behaves like an atomic wavefunction u_{nl}, if $\alpha^2=2|\varepsilon_{nl}^{at}|$, and for small r, if $\frac{\alpha(l+1)}{r}\approx v^{at}(r)$. Together with (2.5) this makes it clear that the expressions

$$b_{l\nu\alpha}(r) = r^{l+\nu-1}\, e^{-\alpha r} \;, \quad \nu = 1,2,\ldots, \quad \alpha = A,\ldots,|\varepsilon_{nl}^{at}|$$

form an appropriate basis for atomic wavefunctions to be expanded into. Our experience is that

$$\boxed{\begin{aligned} b_{l\nu\alpha}(r) &= r^{l+\nu-1}\, e^{-\alpha r} \equiv \langle r|l\nu\alpha\rangle, \\ \nu &= 1,2,3; \quad \alpha = A, A/q, A/q^2,\ldots,|\varepsilon_{nl}^{at}|; \quad q\approx 3 \end{aligned}} \tag{2.10}$$

does it very well, the number of basis functions thereby not exceeding 15.

For the sake of convenience we use in the following Dirac's notation of states. The function $b_{l\nu\alpha}(r)$ is expressed as $\langle r|l\nu\alpha\rangle$, and $\langle l\nu\alpha|l\nu'\alpha'\rangle_w$ means a scalar product

$$\langle l\nu\alpha|l\nu'\alpha'\rangle_w \equiv \int_0^\infty dr\; w(r)\; b_{l\nu\alpha}(r)\; b_{l\nu'\alpha'}(r) \tag{2.11}$$

with an arbitrarily fixed weighting factor w(r), especially

$$\langle l\nu\alpha|l\nu'\alpha'\rangle \equiv \langle l\nu\alpha|l\nu'\alpha'\rangle_{r^2}. \tag{2.12}$$

$|l\nu\alpha\rangle$ will sometimes be abreviated by $|t\rangle$. If

$$\psi(r) \approx \sum_t c_t\, b_t(r) \tag{2.13}$$

is an expansion of a function ψ into the basis $|t\rangle$ which should minimize the integral

$$I \equiv \int_0^\infty dr\; w(r) \left[\psi(r) - \sum_t c_t\, b_t(r)\right]^2 \Longrightarrow \min, \tag{2.14}$$

then

$$\frac{\partial I}{\partial c_t} = 0 : \quad \sum_{t'} c_{t'} \langle t'|t\rangle_w = \langle\psi|t\rangle_w \tag{2.15}$$

provides the linear equations determining the expansion coefficients c_t. The factor $w(r)$ may be used to weight a certain r-range where the expansion should reproduce the function ψ especially well. Due to the appearance of different exponentials in the non-normalized basis (2.10) the normalization integrals $<t|t>$ may differ by several orders of magnitude. This makes the linear equation system (2.15) numerically badly conditioned. It is therefore preferable to work with a normalized basis

$$|t>>_w = n_t^w|t>, \tag{2.16}$$

where n_t^w is the normalizing factor corresponding to the scalar product $<|>_w$. From (2.15) one has now the well conditioned equation system

$$\sum_{t'} (c_{t'} / n_{t'}^w)<<t'|t>>_w = <\psi|t>>_w \tag{2.17}$$

for the coefficients c_t of (2.13). The matrix of this linear equation system, as a matrix of scalar products, is positive definite and hence may easily be decomposed into a product of an upper triangular matrix $\mathbf{D}$ with its transposed,

$$B_{t't} = <<t'|t>>_w = (\mathbf{D}^T\mathbf{D})_{t't}, \tag{2.18}$$

by Cholesky's procedure which may be found in any computer programme library of linear algebra. (For details see e.g. [119].) The solution of (2.17) is then explicitly obtained as

$$c_t = \sum_{t'} <\psi|t'> n_{t'}^w (\mathbf{D}^{-1}\mathbf{D}^{-1T})_{t't}\, n_t^w , \tag{2.19}$$

whereat the inversion of a triangular matrix $\mathbf{D}$ again is a fast procedure.

In this way e.g. the atomic potential $v^{at}(r)$ may be expanded in terms of (2.10):

$$v^{at}(r) \equiv -\frac{1}{r}\, z(r) \approx -\frac{1}{r}\sum_{\nu\alpha} v_{\nu\alpha}\, r^{\nu-1}e^{-\alpha r} \equiv -\frac{1}{r}\sum_{t} v_t\, b_t(r). \tag{2.20}$$

However, as the potential is singular at r=0, it is our experience that it is preferable to fix the value of the potential expansion for r=0 at the exact limit by the constraint

$$A = \sum_{t} v_t\, b_t(0) \equiv \sum_{t} v_t\, \beta_t\ . \tag{2.21}$$

This is easily achieved by modifying (2.14) into

$$\int_0^\infty dr\ w(r)\left[z - \sum_t v_t b_t\right]^2 - 2\lambda \sum_t v_t \beta_t \implies \min, \tag{2.22}$$

where λ is a Lagrange multiplier. Instead of (2.17), we have now

$$\sum_{t'} (v_{t'} / n^W_{t'}) \langle\langle t'|t\rangle\rangle_W = \langle z|t\rangle\rangle_W + \lambda n^W_t \beta_t \tag{2.23}$$

with the explicit solution

$$v_t = \sum_{t'} \left[\langle z|t'\rangle_W + \lambda\beta_{t'}\right] n^W_{t'}\ (D^{-1}D^{-1T})_{t't}\ n^W_t\ . \tag{2.24}$$

Inserting this into (2.21), λ is obtained to be

$$\lambda = \frac{A - \sum_{tt'} \langle z|t'\rangle_W n^W_{t'} (D^{-1})_{t't}\, \gamma_t}{\sum_t \gamma_t^2}\ , \quad \gamma_t = \sum_{t'} (D^{T-1})_{tt'}\ n^W_{t'}\ \beta_{t'}\ . \tag{2.25}$$

Finally, using in (2.7) the wavefunction expansion

$$\varphi_{nl}(r) \approx \sum_{\nu\alpha} a^{(n)}_{\nu\alpha} \langle r|l\nu\alpha\rangle = \sum_t (a^{(n)}_t / n_t) \langle r|t\rangle\rangle \tag{2.26}$$

and multiplying the resulting equation from the left with $\langle\langle t'|$ we have

$$\sum_t \langle\langle t'| \left[-\frac{1}{2r}\frac{d^2}{dr^2} r + \frac{l(l+1)}{2r^2} + v^{at}(r) - \varepsilon^{at}_{nl}\right] |t\rangle\rangle (a^{(n)}_t / n_t) = 0. \tag{2.27}$$

With the same decomposition as in (2.18) and the definition of new coefficients $\bar{a}^{(n)}_t$ by

$$a^{(n)}_t = n_t \sum_{t'} (D^{-1})_{tt'}\ \bar{a}^{(n)}_{t'}\ , \tag{2.28}$$

from (2.27) the ordinary matrix eigenvalue problem

$$\sum_{t} \left[\bar{H}_{t't} - \delta_{t't}\, \varepsilon^{at}_{nl} \right] \bar{a}^{(n)}_{t} = 0 \qquad (2.29)$$

is obtained, where

$$\bar{\mathbf{H}} = \mathbf{D}^{-1T} \,[[\, \langle\langle t'| -\frac{1}{2r}\frac{d^2}{dr^2} r + \frac{l(l+1)}{2r^2} + v^{at}(r)\, |t\rangle\rangle \,]]\, \mathbf{D}^{-1}. \qquad (2.30)$$

($[[A_{t't}]]$ denotes the matrix $\mathbf{A}$ with elements $A_{t't}$.) Note that (2.28) just means the transition from coefficients $a^{(n)}_{t}$ with respect to the non-orthogonal basis $|t\rangle$ to coefficients $\bar{a}^{(n)}_{t}$ with respect to an ortho-normalized basis.

Using m basis functions in (2.26) we end up in (2.29) with an mxm matrix eigenvalue problem, the solution of wich gives approximate values for the first m atomic energy levels with angular momentum l (eigenvalues of (2.7)) as well as the expansion coefficients for the corresponding wavefunctions. Recall that the minimum property of the ground-state solution of Schrödinger's equation guarantees that the lowest eigenvalue of (2.29) is above the lowest energy level of (2.7). An analogous statement also holds for the error sign of the higher eigenvalues (see Appendix 2). Our experience with the basis (2.10) (with ε^{at}_{nl} replaced by roughly the energy value of the highest occupied atomic level) is that it reproduces correctly the leading four to five decimal figures of the energy values of all occupied atomic levels throughout the periodic table. (Of course, the highest levels of (2.29), that is those with $n\approx m$, deviate heavily from the corresponding levels of (2.7) due to the incompleteness of the finite basis (2.10).)

We give now the explicit formulas for the matrix elements of (2.30). First the matrix of the scalar products of basis states is

$$\langle\langle l\nu\alpha | l\nu'\alpha'\rangle\rangle \equiv B^{(l)}_{\nu\alpha;\nu'\alpha'} = B^{(l)}_{\nu'\alpha';\nu\alpha} =$$

$$= \left(\frac{2\alpha}{\alpha+\alpha'}\right)^{\nu+l+1/2} \left(\frac{2\alpha'}{\alpha+\alpha'}\right)^{\nu'+l+1/2} \sqrt{\prod_{i=1}^{\nu'-\nu} \frac{2\nu+2l+i}{\nu+\nu'+2l+i}}\,, \qquad (2.31)$$

where the last expression is correct for the case $\nu'\geq\nu$. (For $\nu'=\nu$ the last factor is equal to one.) It may recursively be calculated according to

$$B^{(l)}_{\nu\alpha;\nu'\alpha'} = B^{(l)}_{\nu\alpha;\nu'-1\alpha'} \frac{2\alpha'}{\alpha+\alpha'} \frac{\nu+\nu'+2l}{[(2\nu'+2l)(2\nu'+2l+1)]^{1/2}} \qquad (2.32)$$

from the $\nu=\nu'$ matrix elements. From the matrix $B^{(1)}$ the D matrix is obtained according to (2.18) and then inverted. Finally, using the potential representation (2.20), the Hamiltonian matrix is

$$<<l\nu\alpha| -\frac{1}{2r}\frac{d^2}{dr^2}r + \frac{l(l+1)}{2r^2} + v^{at}(r)\ |l\nu'\alpha'>> \equiv H^{(1)}_{\nu\alpha;\nu'\alpha'} =$$

$$= \left[\frac{(\nu+l)(\nu'+l)+l(l+1)}{2(\nu+\nu'+2l)(\nu+\nu'+2l+1)}(\alpha+\alpha')^2 - \frac{\alpha(\nu'+l)+\alpha'(\nu+l)}{2(\nu+\nu'+2l)}(\alpha+\alpha') + \frac{\alpha\alpha'}{2} -\right.$$

$$\left. - \frac{\alpha+\alpha'}{\nu+\nu'+2l}\sum_\beta\left(1+\frac{\beta}{\alpha+\alpha'}\right)^{-\nu-\nu'-2l}\sum_\mu\frac{(\nu+\nu'+2l+\mu-1)!}{(\nu+\nu'+2l-1)!}\frac{v_{\mu\beta}}{(\alpha+\alpha'+\beta)^{\mu-1}}\right]*$$

$$* B^{(1)}_{\nu\alpha;\nu'\alpha'} . \qquad (2.33)$$

Of course, these formulas are very easily obtained from the definitions (2.11), (2.12) and (2.16). However, care is needed to write them down in numerically stable expressions as in (2.31) to (2.33).

The only point left open is the choice of the appropriate weighting factor $w(r)$ in the potential expansion. This question may be uniquely answered in the following manner [1]:

Let $v_m(r)$ be an approximate expansion of $v(r)$ in a finite basis of m functions with the property

$$<v - v_m|v - v_m>_w \Longrightarrow \min, \qquad (2.34)$$

and let $\varphi_m(r)$ be the analogous approximation of $\varphi(r)$. Then

$$<v - v_m|\varphi>_w = <v - v_m|\varphi - \varphi_m>_w . \qquad (2.35)$$

The solution of (2.29) provides the approximation of $\varphi_{nl}(r)$ of (2.7) corresponding to $w=r^2$ (this special weight was used in (2.26) because it gives the best approximation of the ground-state level of (2.7) due to the minimum property of Schrödinger's equation). Hence, $<v-v_m|\varphi>_{r^2} \equiv$ $\equiv <v-v_m|\varphi>$ is of second order in the approximation. This means that $r^2\cdot$ $\cdot[v(r) - v_m(r)]\varphi(r)$ is either small or a rapidely oscillating function. In both cases, $<v-v_m|\varphi^2>$ being the potential energy error in the state φ due to the approximation of v by v_m is to be expected small. From (2.20),

1) R. Hayn and H. Eschrig, 1981, unpublished. See Appendix 2 for a proof.

$$\langle v - v_m | v - v_m \rangle = \langle z - z_m | z - z_m \rangle_{w\equiv 1} ,$$

and hence $w\equiv 1$ must be used in (2.22) to (2.25).

After having provided the complete numerics for the problem of a single atom we may now go over to atomic arrangements.

2.2. The LCAO secular equations

In this section the general strategy for finding a variational solution of (2.3) is outlined where $v(\mathbf{r})$ is the (for the moment assumed given) effective potential (2.1) corresponding to a given geometrical arrangement of atoms. With the properties of atomic wavefunctions in mind as they were shown in fig. 2.1 and discussed in the text of the last section, it is quite natural to try to find the solutions of (2.3) in the form

$$|\nu) = \sum_{\mathbf{R}i} |\mathbf{R}i) a_{\mathbf{R}i;\nu} , \tag{2.36}$$

where $|\nu)$ means the ν-th eigenstate of the Hamiltonian $\hat{h}$ of (2.3), $|\mathbf{R}i)$ is the i-th atom-like state (2.6) corresponding to the atomic potential well centred at site $\mathbf{R}$, and $a_{\mathbf{R}i;\nu}$ are the expansion coefficients yet to be determined. The ansatz (2.36) is at least in the inner part of the atoms sufficiently complete as was demonstrated in the last section. Unfortunately this is however generally not the case in the overlap region between the atoms. In order to have a proper representation of the state $|\nu)$ is this region the atom-like wavefunctions $(\mathbf{r}|\mathbf{R}i)$ must be modified there in dependence of the geometrical arrangement of the atoms. This will be investigated in detail in the next chapter for the case of a crystal lattice. For the moment it suffices to assume that the $|\mathbf{R}i)$ are appropriately modified atom-like states. An alternative would be to add to (2.36) a linear combination of plane waves [21,17] which is, however, much less effective.

Next, the atom-like states $|\mathbf{R}i)$ are classified to come in two varieties: core states $|\mathbf{R}c)$ corresponding to inner atomic shells and valence states $|\mathbf{R}n)$. The core wavefunctions of different atoms do not overlap neither do core wavefunctions feel the neighbouring potential wells. Hence, the core states readily are eigenstates of (2.3)[1] with

---

[1] A remark should be made here in order to avoid misunderstandings: The effective potential $v(\mathbf{r})$ of (2.3) is thought of being a superposition of atom-like potential wells at positions $\mathbf{R}$. These potential wells

(cont. on p. 39)

energy eigenvalues $\epsilon_{\mathbf{R}c}$:

$$\hat{h}|\mathbf{R}c) = |\mathbf{R}c)\epsilon_{\mathbf{R}c} \; . \tag{2.37}$$

This circumstance enables us to eliminate from the expression (2.36) for a valence state $|\nu)$ all coefficients $a_{\mathbf{R}c;\nu}$: Since all eigenstates of a given Hamiltonian are orthogonal to each other, any valence state $|\nu)$ must be orthogonal to all core states $|\mathbf{R}c)$. Therefore, instead of (2.36), we may expand $|\nu)$ in a basis of new local valence states $|\mathbf{R}n>$ (denoted by an angular bracket), which have already been orthogonalized to all core states of all atoms:

$$|\mathbf{R}n> = |\mathbf{R}n) - \sum_{\mathbf{R}'c} |\mathbf{R}'c)(\mathbf{R}'c|\mathbf{R}n), \tag{2.38}$$

$$|\nu) = \sum_{\mathbf{R}n} |\mathbf{R}n> c_{\mathbf{R}n;\nu} \; . \tag{2.39}$$

It is quite unimportant that the states (2.38) are no longer normalized: anyhow they are not orthogonal to each other.

Substituting now (2.39) into (2.3) and projecting this equation onto a state $<\mathbf{R}'m|$, the basic LCAO equations are obtained:

$$\sum_{\mathbf{R}n} \left[<\mathbf{R}'m|\hat{h}|\mathbf{R}n> - \epsilon_\nu <\mathbf{R}'m|\mathbf{R}n>\right] c_{\mathbf{R}n;\nu} = 0. \tag{2.40}$$

From the orthonormality of all core states,

$$(\mathbf{R}'c'|\mathbf{R}c) = \delta_{\mathbf{R}'\mathbf{R}}\, \delta_{c'c} \; , \tag{2.41}$$

and with (2.37) we find

$$<\mathbf{R}'m|\hat{h}|\mathbf{R}n> = (\mathbf{R}'m|\hat{h}|\mathbf{R}n) - \sum_{\mathbf{R}''c} (\mathbf{R}'m|\mathbf{R}''c)\epsilon_{\mathbf{R}''c}(\mathbf{R}''c|\mathbf{R}n),$$

$$<\mathbf{R}'m|\mathbf{R}n> = (\mathbf{R}'m|\mathbf{R}n) - \sum_{\mathbf{R}''c} (\mathbf{R}'m|\mathbf{R}''c)(\mathbf{R}''c|\mathbf{R}'n). \tag{2.42}$$

(cont. from p. 38)
may, however, be different from the effective one-particle potentials of isolated atoms due to modifications of the valence charge density. (Particularly they do not have the $-(1/r)$ tails of (2.5).) Such modifications in the core region cause via $v(\mathbf{r})$ also changes of the atomic core states called core relaxations. The atom-like states in (2.36) should be calculated from the modified atom-like potential wells, and therefore the core states $|\mathbf{R}c)$ there are already relaxed.

Inserting this into (2.40) the final equations are

$$\sum_{\mathbf{R}n} (\mathbf{R}'m|\hat{h}^{ps}(\varepsilon_\nu) - \varepsilon_\nu|\mathbf{R}n)c_{\mathbf{R}n;\nu} = 0, \tag{2.43}$$

where $\hat{h}^{ps}$ is the pseudo-Hamiltonian containing the ordinary Phillips-Kleinman pseudo-potential [103]:

$$\hat{h}^{ps}(\varepsilon_\nu) \equiv \hat{h} + \sum_{\mathbf{R}c} |\mathbf{R}c)(\varepsilon_\nu - \varepsilon_{\mathbf{R}c})(\mathbf{R}c| . \tag{2.44}$$

The method of solving (2.43) is the same as for (2.27), i.e. via a Cholesky decomposition of the overlap matrix

$$B_{\mathbf{R}'m;\mathbf{R}n} = (\mathbf{R}'m|\mathbf{R}n) - \sum_{\mathbf{R}''c} (\mathbf{R}'m|\mathbf{R}''c)(\mathbf{R}''c|\mathbf{R}n). \tag{2.45}$$

At this point some comments should be made regarding the equations (2.43). Practically in all applications of this approach, the atom-like wavefunctions $(\mathbf{r}|\mathbf{R}i)$ are expressed in terms of a basis of elementary functions, either of type (2.10) or Gaussians $r^l\exp(-\gamma r^2)$ [40,109]. The natural question arises why not expand the function $(\mathbf{r}|\nu)$ of (2.36) directly in terms of these elementary functions thereby gaining an enlarged variational freedom against (2.39). This has indeed been done or partly been done [20] eventually including some plane waves in addition [21]. The gain, however, at least against the LCAO version described below is very little if any. The prize paid, on the other hand, is a remarkable increase in the algebraic order of the problem (2.43) and a certain loss in the direct interpretability of the results in chemical terms just making LCAO representations so attractive.

Another comment regards further approximations made when computing the matrix elements of (2.43). According to the decay of the wavefunctions $(\mathbf{r}|\mathbf{R}n)$ the matrix elements of (2.43) decay exponentially with the distance $|\mathbf{R}-\mathbf{R}'|$, and so do the matrix elements of the true Hamiltonian $\hat{h}$ in (2.40). But the exponent in the tail of atom-like wavefunctions is rather small, and correspondingly a very poor convergence of the matrix elements was reported in early applications [81]. For instance in the case of Lithium about 25 neighbouring spheres had to be included. This implied a vast number of multi-centre integrals, when the potential of $\hat{h}$ is decomposed into a lattice sum. In this respect the replacement of the Hamiltonian in (2.40) by a pseudo-Hamiltonian in (2.43) is an important gain. Consider a pseudo-potential matrix element $(\mathbf{R}'m|v^{ps}_{\mathbf{R}''}|\mathbf{R}n)$ with $\mathbf{R}''\neq\mathbf{R}'\neq\mathbf{R}\neq\mathbf{R}''$ (three-centre integral). Here the pseudo-potential acts on the smooth tails of both wavefunctions and, as is well known from pseudo-potential theory, the resulting matrix element is rather small compared to the true potential matrix element $(\mathbf{R}'m|v_{\mathbf{R}''}|\mathbf{R}n)$, in the given case demonstrating a large cancellation between the three-centre potential matrix elements and the core orthogonalization corrections. This improves the convergence of (2.43) against (2.40) already remarkably. A further improvement is obtained by using compressed basis functions as described in the next chapter. In earlier applications the core orthogonalization corrections often have been neglected. Our discussion shows that such a approximation would make sense only if at the same time the three-centre matrix elements would be dropped.

The energy values of the finite algebraic eigenvalue problem (2.43) have again the property that they approximate the corresponding eigenvalues of (2.3) from above. This general feature of variational methods provides a very useful check in numerical calculations.

2.3. The case of the bandstructure of a crystalline solid

From now on the case is treated where the geometrical arrangement of atoms is an infinite regular crystal lattice. First we introduce a notation used troughout the following text:

$$\mathbf{R} = l\mathbf{a}_1 + m\mathbf{a}_2 + n\mathbf{a}_3 \tag{2.46}$$

is a lattice vector and $\mathbf{S}$ a relative atom position vector within the unit cell of the lattice, so that the atom sites are $\mathbf{R}+\mathbf{S}$ (fig. 2.2). The lattice vector of the reciprocal lattice is denoted by $\mathbf{G}$:

$$\mathbf{RG} = 2\pi * \text{integer} \qquad \text{for all } \mathbf{R}. \tag{2.47}$$

The number of unit cells of the lattice is N, tending to infinity. As long as N is finite, cyclic boundary conditions (see e.g. [122]) are assumed in order to maintain the full discrete translational invariance $f(\mathbf{r})=f(\mathbf{r}+\mathbf{R})$ for all $\mathbf{R}$, obeyed by any observable quantity f. Such a quantity is for instance the square of the wavefunction,

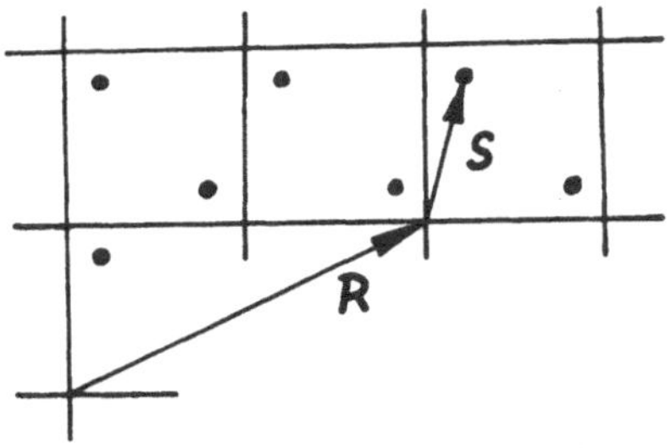

Fig. 2.2. Lattice vector **R** and relative atom position **S**.

and it is the assertion of Bloch's theorem [48], that the solutions of (2.3) for a potential v with lattice periodicity may be chosen to fulfil Bloch's condition

$$\varphi_{\mathbf{k}\nu}(\mathbf{r}) = e^{i\mathbf{kR}} \varphi_{\mathbf{k}\nu}(\mathbf{r}-\mathbf{R}), \tag{2.48}$$

where $\mathbf{k}$ mod $\mathbf{G}$ is the quasi-momentum of the electron in the state $\varphi_{\mathbf{k}\nu}$ (hence, any quantity depending on the quasi-momentum is a periodic function of $\mathbf{k}$ with the periods of the reciprocal lattice). Summation over all $\mathbf{k}$ means [122]

$$\sum_{\mathbf{k}} = N \frac{V_{U.C.}}{(2\pi)^3} \int_{B.Z.} d^3k \, , \tag{2.49}$$

where $V_{U.C.}$ is the volume of the unit cell of the lattice and the integration is over the first Brillouin zone (B.Z.), the entirely symmetric unit cell of the reciprocal lattice. (Numerical procedures for carrying out the $\mathbf{k}$-space integrations are described in chapter 6.)

For any given $\mathbf{k}$ a state with the property (2.48) is easily constructed from the atom-like states $|\mathbf{R}+\mathbf{S}n)$ or their variants (2.38)[1]:

$$\boxed{|\mathbf{kS}n\rangle = \frac{1}{\sqrt{N}} \sum_{\mathbf{R}} e^{i\mathbf{k}(\mathbf{R}+\mathbf{S})} \, |\mathbf{R}+\mathbf{S}n\rangle .} \tag{2.50}$$

Though the state $|\mathbf{kS}n\rangle$ is not normalized (due to the non-orthogonality of the atom-like states), the prefactor of the Bloch sum on the right hand side of (2.50) ensures that its normalization integral does not depend on the crystal size. Inversely, only linear combinations of states (2.50) with the same $\mathbf{k}$ obey the condition (2.48). Hence, the appropriate trial function for the Bloch state (2.48) is

$$\boxed{|\mathbf{k}\nu\rangle = \sum_{\mathbf{S}n} |\mathbf{kS}n\rangle C_{\mathbf{S}n;\nu}(\mathbf{k})} \tag{2.51}$$

instead of (2.39), leading after a projection onto $e^{-i\mathbf{kS}'}\langle\mathbf{S}'m|$ to the equations

$$\boxed{\sum_{\mathbf{S}n} \left[\sum_{\mathbf{R}} e^{i\mathbf{k}(\mathbf{R}+\mathbf{S}-\mathbf{S}')} (\mathbf{S}'m|\hat{h}^{ps}(\varepsilon_\nu(\mathbf{k})) - \varepsilon_\nu(\mathbf{k})|\mathbf{R}+\mathbf{S}n) \right] C_{\mathbf{S}n;\nu}(\mathbf{k}) = 0} \tag{2.52}$$

1) The additional constant phase factor exp($i\mathbf{kS}$) is introduced just for convencience.

instead of (2.43). They are now solved by Cholesky-decomposing the **k**-dependent overlap matrix

$$B_{\mathbf{S}'m;\mathbf{S}n}(\mathbf{k}) = \sum_{\mathbf{R}} e^{i\mathbf{k}(\mathbf{R}+\mathbf{S}-\mathbf{S}')} \, B_{\mathbf{S}'m;\mathbf{R}+\mathbf{S}n} \, , \tag{2.53}$$

constructed from the expressions (2.45), into the product $\mathbf{D}^T\mathbf{D}$ and then diagonalizing the Hamiltonian matrix

$$\bar{H}(\mathbf{k}) = \mathbf{D}^{-1T}(\mathbf{k}) \, H(\mathbf{k}) \, D^{-1}(\mathbf{k}), \tag{2.54}$$

where

$$H_{\mathbf{S}'m;\mathbf{S}n}(\mathbf{k}) = \sum_{\mathbf{R}} e^{i\mathbf{k}(\mathbf{R}+\mathbf{S}-\mathbf{S}')} \, (\mathbf{S}'m| \left[\hat{h} - \sum_{\mathbf{R}''\mathbf{S}''c} |\mathbf{R}''\mathbf{S}''c)\varepsilon_{\mathbf{S}''c}(\mathbf{R}''\mathbf{S}''c| \right] |\mathbf{R}+\mathbf{S}n). \tag{2.55}$$

In the general case, the matrices **B(k)** and **H(k)** are complex (Hermitean). If, however, the crystal lattice has a centre of inversion, then these matrices may be directly transformed into real ones. As this transformation is of practical importance, it is described here. There may be several centres of inversion within te unit cell of a lattice (fig. 2.3), but their distribution is strongly correlated with the translational symmetry: Be **I** and **I'** two different centres of inversion, be $\hat{I}$ and $\hat{I}'$ the corresponding operations of inversion, i.e., $\hat{I}(\mathbf{r}-\mathbf{I})=\mathbf{I}-\mathbf{r}$, and be $\hat{T}_{\mathbf{R}}$ a translation by the vector **R**. Then it is easy to see that $\hat{I}'\hat{I}=T_{2(\mathbf{I}'-\mathbf{I})}$. In other words, $2(\mathbf{I}'-\mathbf{I})$ is necessarily a lattice vector for any pair of centres of inversion **I** and **I'**. Inversly, if **I** is a centre of inversion, **R** any lattice vector, and $\mathbf{I}'=\mathbf{I}-\mathbf{R}/2$, then the relation $\hat{I}\hat{T}_{\mathbf{R}}=\hat{I}\hat{I}\hat{I}'=\hat{I}'$ shows that **I'** is also a centre of

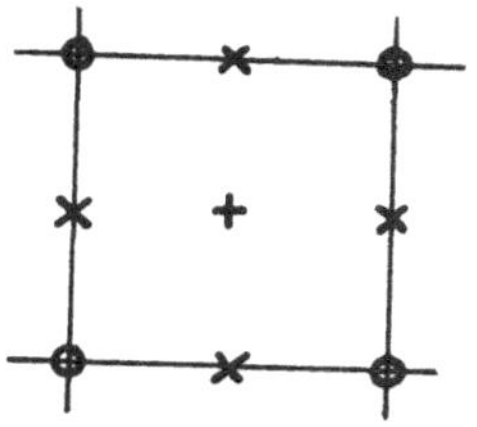

Fig. 2.3. The symbols o, +, × denote different centres of inversion of a twodimensional rectangular Bravais lattice. The distance vector of any two centres of inversion is half a lattice vector.

inversion. Let now **S** be any vector. Consider the chain of equalities

$$\hat{I}\mathbf{S} = \hat{I}\hat{T}_{2(\mathbf{I}-\mathbf{I}')}\,(\mathbf{S} - 2(\mathbf{I}-\mathbf{I}')) = \hat{I}^2\hat{I}'(\mathbf{S} - 2(\mathbf{I}-\mathbf{I}')) = \hat{I}'\mathbf{S} - \mathbf{R},$$

where $\mathbf{R}=\hat{I}'2(\mathbf{I}-\mathbf{I}')=2(\mathbf{I}'-\mathbf{I})$ is a lattice vector. It proves that for any vector **S** all vectors $\hat{I}\mathbf{S}$ corresponding to all centres of inversion **I** are translationally equivalent to each other.

Choose now any centre of inversion **I** as the origin of the lattice and construct the Wigner-Seitz cell around it, i.e. the unit cell which has the full point symmetry of the Bravais lattice [122]. Then all atom positions **S** come in two classes: atoms at centres of inversion for which $\hat{I}\mathbf{S}=-\mathbf{S}=\mathbf{S}-\mathbf{R}$ already belongs to the next unit cell (except for $\mathbf{S}=0$), and atoms not at centres of inversion which come always in pairs **S** and $\hat{I}\mathbf{S}=-\mathbf{S}$ translationally non-equivalent and therefore belonging to the same Wigner-Seitz cell. This classification (inclusive the grouping in pairs) is obviously independent of the chosen centre **I**. Number the atom sites within the unit cell in such a way that the first n_I sites $\mathbf{S}_i$, $i=1,\dots,n_I$ are those at centres of inversion and that $\hat{I}\mathbf{S}_i=-\mathbf{S}_i=\mathbf{S}_{i+n_A}$ for $i=n_I+1,\dots,n_I+n_A$, the total number of atoms in the cell being n_I+2n_A. The atom-like wavefunctions (2.6) with a real radial part φ_{nl} and with the spherical harmonics of Appendix 1 have the transformation property $\langle-\mathbf{r}|-\mathbf{R}n\rangle = \langle\mathbf{r}|\mathbf{R}n\rangle^*$, hence

$$\langle-\mathbf{R}'-\mathbf{S}'m|\hat{h}|-\mathbf{R}-\mathbf{S}n\rangle = \langle\mathbf{R}'+\mathbf{S}'m|\hat{h}|\mathbf{R}+\mathbf{S}n\rangle^* = \langle\mathbf{R}+\mathbf{S}n|\hat{h}|\mathbf{R}'+\mathbf{S}'m\rangle \tag{2.56}$$

for any Hermitean $\hat{h}$ with $[\hat{h},\hat{I}]=0$. Consequently,

$$\begin{aligned}\langle\mathbf{k},-\mathbf{S}'m|\hat{h}|\mathbf{k},-\mathbf{S}n\rangle &= \sum_{\mathbf{R}} e^{i\mathbf{k}(\mathbf{R}+\mathbf{S}'-\mathbf{S})}\,\langle-\mathbf{S}'m|\hat{h}|\mathbf{R}-\mathbf{S}n\rangle = \\ &= \sum_{\mathbf{R}} e^{i\mathbf{k}(\mathbf{R}+\mathbf{S}'-\mathbf{S})}\,\langle\mathbf{S}'m|\hat{h}|-\mathbf{R}+\mathbf{S}n\rangle^* = \\ &= \sum_{\mathbf{R}} e^{-i\mathbf{k}(\mathbf{R}+\mathbf{S}-\mathbf{S}')}\,\langle\mathbf{S}'m|\hat{h}|\mathbf{R}+\mathbf{S}n\rangle^* = \\ &= \langle\mathbf{k}\mathbf{S}'m|\hat{h}|\mathbf{k}\mathbf{S}n\rangle^*.\end{aligned}$$

Together with our special numbering of atom sites this implies for (2.55)

$$H_{\mathbf{S}_{i'}m;\mathbf{S}_{j'}n}(\mathbf{k}) = (H_{\mathbf{S}_i m;\mathbf{S}_j n}(\mathbf{k}))^*, \tag{2.57}$$

where

$$\begin{aligned} i' &= i && \text{for} && i = 1,\dots,n_I, \\ i' &= i+n_A && \text{for} && i = n_I+1,\dots,n_I+n_A\,, \\ i' &= i-n_A && \text{for} && i = n_I+n_A+1,\dots,n_I+2n_A. \end{aligned}$$

Analogous relations are true for the **B**-matrix of (2.53).

If we introduce the block matrix notation

$$\mathbf{H}(\mathbf{k}) = \begin{pmatrix} \mathbf{H}_{II} & \mathbf{H}_{IA} & \mathbf{H}_{IA'} \\ \mathbf{H}_{AI} & \mathbf{H}_{AA} & \mathbf{H}_{AA'} \\ \mathbf{H}_{A'I} & \mathbf{H}_{A'A} & \mathbf{H}_{A'A'} \end{pmatrix} \tag{2.58}$$

evident by itself, then the inversion symmetry implies

$$\mathbf{H}(\mathbf{k}) = \begin{pmatrix} \mathbf{H}_{II} & \mathbf{H}_{IA} & \mathbf{H}^*_{IA} \\ \mathbf{H}^+_{IA} & \mathbf{H}_{AA} & \mathbf{H}_{AA'} \\ \mathbf{H}^T_{IA} & \mathbf{H}^*_{AA'} & \mathbf{H}^*_{AA} \end{pmatrix} = \mathbf{H}^+(\mathbf{k}), \tag{2.59}$$

where

$$\mathbf{H}_{II} = \mathbf{H}^*_{II} = \mathbf{H}^T_{II}, \quad \mathbf{H}_{AA} = \mathbf{H}^+_{AA}, \quad \mathbf{H}_{AA'} = \mathbf{H}^T_{AA'}, \tag{2.60}$$

and only these three block matrices need be computed.

Denoting by $\mathbf{1}$ a unit block matrix we introduce the unitary matrix

$$\mathbf{U} = \frac{1}{\sqrt{2}} \begin{pmatrix} \sqrt{2}*\mathbf{1} & 0 & 0 \\ 0 & \mathbf{1} & i*\mathbf{1} \\ 0 & \mathbf{1} & -i*\mathbf{1} \end{pmatrix}, \quad \mathbf{U}^{-1} = \mathbf{U}^+. \tag{2.61}$$

It is now straightforward to see, that

$$\tilde{\mathbf{H}}(\mathbf{k}) \equiv \mathbf{U}^+\mathbf{H}(\mathbf{k})\mathbf{U} =$$

$$= \begin{pmatrix} \mathbf{H}_{II} & \sqrt{2}*\mathrm{Re}\mathbf{H}_{IA} & -\sqrt{2}*\mathrm{Im}\mathbf{H}_{IA} \\ \sqrt{2}*\mathrm{Re}\mathbf{H}^T_{IA} & \mathrm{Re}\mathbf{H}_{AA}+\mathrm{Re}\mathbf{H}_{AA'} & -\mathrm{Im}\mathbf{H}_{AA}+\mathrm{Im}\mathbf{H}_{AA'} \\ -\sqrt{2}*\mathrm{Im}\mathbf{H}^T_{IA} & \mathrm{Im}\mathbf{H}_{AA}+\mathrm{Im}\mathbf{H}_{AA'} & \mathrm{Re}\mathbf{H}_{AA}-\mathrm{Re}\mathbf{H}_{AA'} \end{pmatrix} \tag{2.62}$$

is already a real symmetric matrix. Analogously, $\mathbf{B}(\mathbf{k})=\mathbf{U}^+\mathbf{B}(\mathbf{k})\mathbf{U}$. On a computer the matrix (2.59) may be stored in a real array according to the pattern

$$\begin{pmatrix} \mathbf{H}_{II} & & \\ \mathrm{Re}\mathbf{H}_{IA} & \mathrm{Re}\mathbf{H}_{AA} & \mathrm{Re}\mathbf{H}_{AA'} \\ \mathrm{Im}\mathbf{H}_{IA} & \mathrm{Im}\mathbf{H}_{AA} & \mathrm{Im}\mathbf{H}_{AA'} \end{pmatrix}.$$

The transformation to (2.62) can than easily be made on place.

Many crystal lattices have only atoms on inversion centres, this is particularly so for any Bravais lattice. In those cases, e.g. the Hamiltonian matrix consists only of $\mathbf{H}_{II}$. It was the main reason for introducing a factor i in the spherical harmonics of Appendix 1 for odd angular momenta l that in this way the LCAO matrices are automatically obtained real for those lattices.

2.4. Multi-centre integrals

There exists a number of different approaches numerically to compute the matrix elements of (2.43) and (2.52), respectively. The periodic $\mathbf{r}$-dependence of the Hamiltonian favours a Fourier expansion of the crystal potential and of the basis wavefunctions [116]. The drawback is a poor convergence due to the singularities of the potential at the atom positions suggesting a mixed treatment handling the singular part in $\mathbf{r}$-space. The widely used alternative is to represent the potential as a lattice sum

$$v(\mathbf{r}) = \sum_{\mathbf{RS}} v_{\mathbf{S}}(\mathbf{r}-\mathbf{R}-\mathbf{S}) \equiv \sum_{\mathbf{RS}} v_{\mathbf{RS}} \tag{2.63}$$

where the potential function $v_{\mathbf{S}}$ at the atom site $\mathbf{R}+\mathbf{S}$ depends only on $\mathbf{S}$ and not on $\mathbf{R}$. This decomposition implies a decomposition of the matrix element

$$\begin{aligned}(\mathbf{S}'m|\hat{h}^{ps}(\varepsilon)-\varepsilon|\mathbf{R}+\mathbf{S}n) &= \\ &= -\varepsilon B_{\mathbf{S}'m;\mathbf{R}+\mathbf{S}n} + (\mathbf{S}'m|\hat{t}+v_{\mathbf{RS}}+(1-\delta_{\mathbf{R}+\mathbf{S},\mathbf{S}'})v_{\mathbf{0S}'}|\mathbf{R}+\mathbf{S}n) + \\ &\quad + {\sum_{\mathbf{R}''\mathbf{S}''}}' (\mathbf{S}'m|v^{ps}_{\mathbf{R}''\mathbf{S}''}|\mathbf{R}+\mathbf{S}n)\end{aligned} \tag{2.64}$$

into two-centre integrals (first line on the right hand side; the $\mathbf{B}$-matrix is that of (2.45)) and three-centre integrals (second line). Due to the fact that by construction the atom-like core and valence states centred at the same atom site are orthogonal to each other and the core states are eigenstates of the crystal Hamiltonian $\hat{h}$ (see (2.37)), the two-centre pseudopotential matrix elements are identical with the corresponding ordinary potential matrix elements. The prime at the lattice sum over the tree-centre terms of (2.64) means that the items with $\mathbf{R}''+\mathbf{S}''=\mathbf{S}'$ and with $\mathbf{R}''+\mathbf{S}''=\mathbf{R}+\mathbf{S}$ are to be omitted. In the case $\mathbf{S}'=\mathbf{R}+\mathbf{S}$ the "degenerate three-centre integrals" of the last line of (2.64) are called crystal field integrals because they describe the crystal field splitting of the levels in the tight-binding model (where the $\mathbf{B}$-matrix is replaced by the unit matrix).

If all radial dependences are expanded into Gaussians, all these integrals may be done analytically [7,6] though the resulting expressions, especially for the case of higher angular momenta, are rather lengthy. In section 2.1 we gave arguments in favour of the use of the basis (2.10) for radial dependences. The particularly appropriate character of these functions allows to keep the total number of basis elements smaller than in the case of Gaussians. The prize is that now only two-centre integrals may be done analytically. In the LCAO version furher described in the next chapter, the number of three-centre integrals to be computed is very limited, and therefore we prefer the basis (2.10) and give here the details how to handle the corresponding integrals.

If potentials and atom-like wavefunctions are expanded in terms of basis functions (sometimes called Slater-type functions)

$$\langle \mathbf{r} | \mathbf{R} l m \nu \alpha \rangle = x^{l+\nu-1} \, e^{-\alpha x} \, Y_{lm}(\hat{\mathbf{x}}) \, , \tag{2.65}$$

where $\mathbf{x}$ is the relative coordinate vector $\mathbf{x}=\mathbf{r}-\mathbf{R}$, then, after reexpanding a product of two such functions at the same centre and after rotating the coordinate system (transition from global to local coordinates with the explicit formulas given in Appendix 1.), any two-centre integral reduces to a linear combination of terms

$$\langle a | b \rangle \equiv \langle 0 l_a m \nu_a \alpha_a | (R \mathbf{e}_z) l_b m \nu_b \alpha_b \rangle \tag{2.66}$$

being only nonzero for $m_a = m_b \equiv m$. The integral (2.66) can be done in spheroidal coordinates [108]. The unpleasant problem with these terms is that the resulting expressions are numerically highly unstable (tending to limits $\infty - \infty$). An atempt to stabilize the expressions by rearranging them in an appropriate manner was made in [52], however, the formulas given there still are unstable in the case $|\alpha_a - \alpha_b| R \gtrsim 10$ for higher numbers ν. A mathematically very elegant treatment to circumvent the difficulties was given in [41] by the use of so-called reduced Bessel functions having very convenient transformation properties. At the prize of another expansion in terms of those Bessel functions all difficulties disappear.

An alternative treatment is described in [32] avoiding another reexpansion and resulting in the rather stable expressions

$$i^{(l_a-l_b)} \langle a|b\rangle = \frac{1}{2}\left(\frac{R}{2}\right)^{l_a+\nu_a+l_b+\nu_b+1} *$$

$$* \sqrt{\frac{(2l_a+1)(2l_b+1)}{(l_a+m)!(l_a-m)!(l_b+m)!(l_b-m)!}} *$$

$$* \sum_{ll'} M_{ll'}^{(ab)} \begin{cases} \dfrac{e^{-\alpha_a R}}{\alpha\beta} f_{l',k-l'}^{(l_b+\nu_b)}(\alpha,\beta) - (-1)^{k-l'} \dfrac{e^{-\alpha_b R}}{\alpha\beta} f_{l',k-l'}^{(l_a+\nu_a)}(\alpha,-\beta) \\ \dfrac{e^{-\alpha_a R}}{\alpha_a R} f_{l'}^{(0)}(\alpha) \dfrac{(-1)^{k-l'+1} - 1}{k - l' + 1} \end{cases} , \tag{2.67}$$

where $k=2l+\nu_a+\nu_b$, $\alpha=(\alpha_a+\alpha_b)R/2$, $\beta=(\alpha_a-\alpha_b)R/2$, and the lower line of (2.67) is for $\beta=0$. The functions f are given by the recurrence formulas

$$f_0^{(k)} = \delta_{k0}\ , \quad f_l^{(k)}(\alpha) = f_{l-1}^{(k)}(\alpha)\,\frac{l}{\alpha} + \frac{l!}{(l-k)!\alpha^k} \equiv f_{l0}^{(k)}(\alpha,\beta)$$

$$f_{l-1,l'+1}^{(k)}(\alpha,\beta) = \frac{\alpha}{\beta}\,\frac{l'+1}{l}\left[f_{ll'}^{(k)}(\alpha,\beta) - f_{l'}^{(k)}(\beta)\right] + f_{l-1}^{(k)}(\alpha)\ . \tag{2.68}$$

They are symmetric according to

$$f_{ll'}^{(k)}(\alpha,\beta) = f_{l'l}^{(k)}(\beta,\alpha)\ . \tag{2.69}$$

The integer coefficients $M_{ll'}^{(ab)}$ are nonzero only for

$$0 \le l \le l_a+l_b,$$

$$\max(0,2l-l_a-l_b) \le l' \le \min(l_a+l_b+\nu_a+\nu_b, 2l+\nu_a+\nu_b). \tag{2.70}$$

For $\nu_a=\nu_b=0$ they may be computed once forever from the explicit expressions

$$M_{ll'}^{(ab)} = \sum_{p=0,2,\ldots}^{l_a-m} \binom{l_a-m}{p} (2l_a-1-p)!!\,(p-1)!! \,*$$

$$* \sum_{q=0,2,\ldots}^{l_b-m} \binom{l_b-m}{q} (2l_b-1-q)!!\,(q-1)!! \,*$$

$$* \sum_{u=0}^{p} \binom{p}{u} \sum_{v=0}^{q} \binom{q}{v} \sum_{u'=0}^{l_a-m-p} \binom{l_a-m-p}{u'} \sum_{v'=l_v}^{l_b-m-q} \binom{l_b-m-q}{v'} \,*$$

$$* \binom{m}{(l'-u-v-u'-v')/2} \binom{m}{(2l-l'-p-q+u+v-u'-v')/2} \,*$$

$$* (-1)^{l_b+l+v+u'+1} , \qquad (2.71)$$

where $l!!$ means the double factorial $1\cdot3\cdots l$ for odd l and $2\cdot4\cdots l$ for even l, $(-1)!!=0!!=1$, and $\binom{m}{l}=0$ for $l<0$ or $l>m$; finally, $l_v=(l'-u-v-u')$ mod 2. For $\nu_a>0$ and/or $\nu_b>0$ the recurrence formulas

$$M_{ll'}^{(ab)} = M_{ll'-1}^{(a\_b)} + M_{ll'}^{(a\_b)} , \quad \nu_a > 0$$

$$M_{ll'}^{(ab)} = M_{ll'-1}^{(ab\_)} - M_{ll'}^{(ab\_)} , \quad \nu_b > 0 \qquad (2.72)$$

may be used where (a_b) and (ab_) mean the cases with ν_a replaced by ν_a-1 and ν_b replaced by ν_b-1, respectively.

The three-centre integrals must be done numerically, preferably in spherical coordinates arround the potential centre. For those readers interested in the numerics we close this chapter with the description of an approved numerical procedure for the three-centre terms.

Let v_0 be a potential well centred at the origin and consider the integral $\langle \mathbf{X}'a|v_0|\mathbf{X}b\rangle$ with $\mathbf{X}'$ and $\mathbf{X}$ being site vectors of the type $\mathbf{R}+\mathbf{S}$ for the two atom-like wavefunctions a and b. Define a local coordinate system in such a way that $X_1'=X_1$ and $X_2'=X_2=0$, that is, the vector $\mathbf{X}'-\mathbf{X}$ is parallel to the local z-axis as in (2.66). The geometry is shown in fig. 2.4. The transition from the global to these local coordinates is exactly the same as in (2.66). Though there is generally no m-selection rule as it was for the two-centre terms, the total contribution of the $m_a\neq m_b$ terms to the LCAO matrices is however generally negligible small in the version with the special atom-like states of the next chapter.

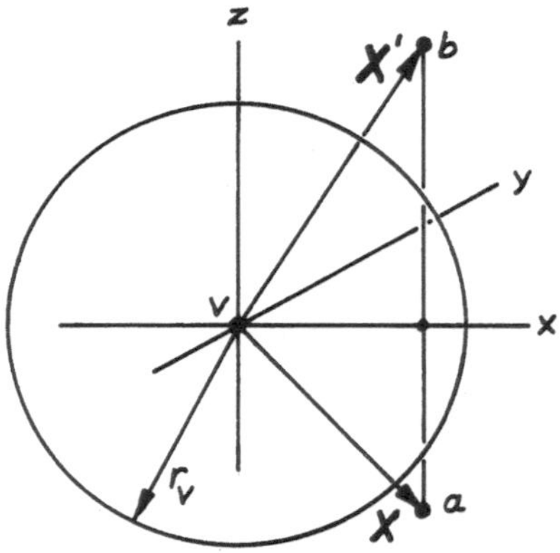

Fig. 2.4. Local coordinates for the three-centre integrals. The range r_v of the potential should be well below min(|**X'**|,|**X**|).

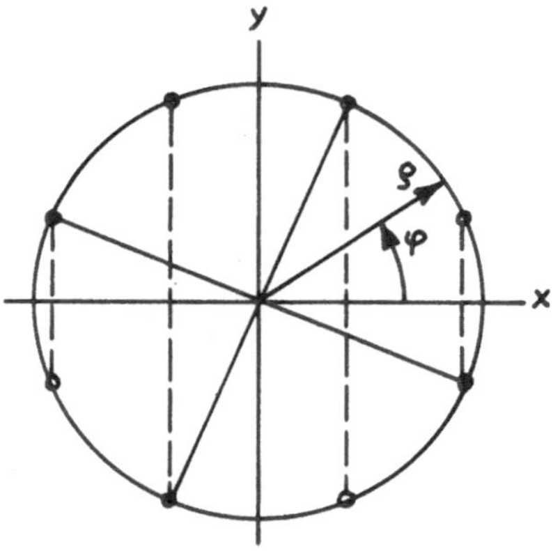

Fig. 2.5. Points for the numerical φ-integration for $N_\varphi=4$.

Spherical coordinates are introduced now with the z-axis as the polar axis, and the first integation is over the azimut φ around this axis From 0 to 2π, where in the following φ=0 corresponds to y=0. If the "real" spherical harmonics of Appendix 1 are used and the geometry is that of fig. 2.4, then the integrand of $\langle \mathbf{X}'a|v_0|\mathbf{X}b\rangle$ will be either symmetric or antisymmetric with respect to a reflection y → -y, the integral in the latter case (for an inversional symmetric v_0) being zero. When this symmetry is present, a φ-integration from 0 to π would do. In order that the routine works in the general case (for any v_0 and arbitrary vectors **X'** and **X**), the points of fig. 2.5 are used in a full integration from 0 to 2π:

$$\varphi_i = \frac{\pi}{N_\varphi}(2i - 1.5),\ i = 1,\ldots,N_\varphi. \tag{2.73}$$

This choice is in the symmetric case equavalent to an integration in N_φ points from 0 to π. The number of points N_φ is chosen proportional to r and a simple equal weight summation is used[1].

[1] It was pointed out to the author many yeas ago by W. A. Harrison that in the case of an integral over the full period of an analytic periodic function equal weight integration is the best one can do: Take any integration formula $\int dx f(x)=\sum W_i f(x_i)$. Apply it succesively to periods of f starting with all points x_i and average the results. (cont. on p. 51)

Next the $\zeta = \cos\theta$-integration is performed again as an equal weight summation with a number of points proportional to r, and finally the r-integral is done using Simpson's formula being favoured by the behaviour of the integral. The result is

$$\langle \mathbf{X}'a|v_0|\mathbf{X}b\rangle = \frac{4\pi r_v^3}{N^3} \sum_{i=1}^{N-1} W_i \sum_{j=1}^{i} \sum_{k=1}^{i} \langle \mathbf{X}'a|\mathbf{r}_{ijk}\rangle v_0(\mathbf{r}_{ijk}) \langle \mathbf{r}_{ijk}|\mathbf{X}b\rangle \tag{2.74}$$

with $\mathbf{r}_{ijk}$ being given in spherical coordinates by

$$\begin{aligned} r_i &= \frac{i}{N} r_v , \\ \zeta_{ji} &= -1 + (2j-1)/i , \\ \varphi_{ki} &= (2k-1.5)\pi/i , \end{aligned} \tag{2.75}$$

and

$$W_i = \begin{cases} 2/3 & \text{for odd } i. \\ 1/3 & \text{for even } i. \end{cases} \tag{2.76}$$

The choice (2.75) of points already accounts for the Jacobian r^2. Fig. 2.6 shows the distribution of points for N=3. Our experience is that N=10 ensures an accuracy of 10^{-4} a.u.

The core orthogonalization corrections at the potential site of fig. 2.4 may be cast into a sum of expressions

$$\sum_{m_c} \langle \mathbf{X}'a|c\rangle\langle c|\mathbf{X}b\rangle =$$

$$= \sum_{m_1 m_2} C^{l_a}_{m_1 m_a} \langle am_1|cm_1\rangle \, \varepsilon^{l_c}_{m_1 m_2} \langle cm_2|bm_2\rangle \, C^{l_b}_{m_2 m_b} , \tag{2.77}$$

(cont. from p. 50)

You are left just with the equal weight result showing that on an average the error of any integration formula is just that of the equal weight formula. The argument does of course not apply to a periodic function with a singularity at x_o because now the period starting with x_o is distinguished.

where l_c, m_c are the angular quantum numbers of a core state $|c\rangle$ and $\langle am_1|cm_1\rangle$ means an integral (2.66) with the core state at the origin and the valence state $|a\rangle$ at $|\mathbf{X}'|\mathbf{e}_z$, and both having an azimut quantum number m_1. The coefficients $C^{l_a}_{m_1 m_a}$ are those appearing when the spherical harmonic of the state $|\mathbf{X}'a\rangle$ is rotated into the new orientation and are again given in Appendix 1. The two-centre integrals $\langle am_1|cm_1\rangle$ contain a factor $(r_c/|\mathbf{X}'|)^m \approx (0.2...0.1)^m$, where r_c is the core radius, and only terms up to first order in this ratio need be taken in (2.77). That implies that all terms with negative m_a, m_b, m_1 or m_2 and with $m_1+m_2>1$ may be neglected. Finally, $\varepsilon^{l_c}_{m_1 m_2}$ is the projection of $|cm_1\rangle$ onto $|cm_2\rangle$ (being differently oriented according to $\mathbf{X}'$ and $\mathbf{X}$). With $e = \cos(\mathbf{X}',\mathbf{X})$ we have

Fig. 2.6. Distribution of points for the three-centre integration with N=3.

$$\varepsilon^0 = 1$$

$$\boldsymbol{\varepsilon}^1 = \begin{pmatrix} e & \sqrt{1-e^2} \\ -\sqrt{1-e^2} & \cdots \end{pmatrix}$$

$$\boldsymbol{\varepsilon}^2 = \begin{pmatrix} (3e^2-1)/2 & \sqrt{3}e\sqrt{1-e^2} \\ -\sqrt{3}e\sqrt{1-e^2} & \cdots \end{pmatrix} \tag{2.78}$$

$$\boldsymbol{\varepsilon}^3 = \begin{pmatrix} (5e^3-3e)/2 & \sqrt{3/8}(5e^2-1)\sqrt{1-e^2} \\ -\sqrt{3/8}(5e^2-1)\sqrt{1-e^2} & \cdots \end{pmatrix}$$

(Terms with $\varepsilon^{l_c}_{11}$ were already neglected.) Of course, instead a local pseudopotential could be used.

3. OPTIMIZATION OF THE BASIS

This chapter, though again of technical character, plays a key role in our treatise, because it is shown here how to convert LCAO into a highly effective and precise procedure, and the ability of this optimized variant of a scheme originally contrived for the tight-binding case is just demonstrated at the opposite extreme: for the free electron limit.

In the first section a parametrization of compressed atom-like basis states is introduced wich allows to find optimum LCAO basis states by another variational procedure described in section 3.2 and depending only on the lattice type but not on the lattice constant and not on the crystal potential. The accuracy of the scheme is illustrated by LCAO representations of the free-electron band structures for various lattice types and, in section 3.3, at the example of the copper band structure. Complete instructions are given for an optimized basis LCAO band calculation for any metallic material (provided a suitable crystal potential is at hand, this problem being treated in chapter 7.)

3.1. Parametrization of the basis states

We learned already in section 2.1 that a basis of one-particle wavefunctions of free atoms is not well suited for an LCAO treatment of broad bands in a crystal. Being quite appropriate in the inner région of the atoms, free atom wavefunctions do not form a sufficiently complete basis in the interstitial region for energies well above the potential floor. Furthermore, the far-reaching smooth tails of neighbouring basis functions even spoil the wavefunction (2.39) in the inner region of a given atom adding there an incorrect nearly constant term to the rapidly oscillating correct contribution of the basis functions centred at that site. These simple considerations show that an atom wavefunction should be somehow compressed before being used as an LCAO basis state. Of course, the atom wavefunction may not simply be cut at some radius because in this way a singularity would be introduced leading to completely incorrect kinetic energy terms. A smooth radial decaying factor would do, but a variational method, as LCAO is, generally cannot be improved just by doing violence to it: the modification of the basis states should again be a variational procedure. Atom-like basis states compressed in a "hand-made" manner were used previously by several authors (see, e.g., [40]). Since anyhow the basis states must be calculated from an equation of type (2.4), the modification containing variational parameters may be incorporated in this equation.

Representing the crystal potential in the form (2.63) the appropriate LCAO basis functions may be obtained from the equation [35]

$$\left[-\frac{\Delta}{2}+v_S(\mathbf{r})+\left(\frac{r}{r_{Sn}}\right)^{n_o}\right]\varphi_{Sn}(\mathbf{r})=\varepsilon_{Sn}\varphi_{Sn}(\mathbf{r}) \tag{3.1}$$

containing the variational parameters r_{Sn} and n_o. The states entering (2.52) to (2.55) are then

$$(\mathbf{r}|\mathbf{R}+Sn)=\varphi_{Sn}(\mathbf{r}-\mathbf{R}-\mathbf{S}).$$

The artifical attractive potential $(r/r_{Sn})^{n_o}$ was introduced in (3.1) as a tool to compress the atom-like wavefunction φ_{Sn} without changing its behaviour inside the atom. Intuitively it is clear that r_{Sn} should be of the order of the atom radius, and that n_o must probably be larger then 2 in order not to perturb φ_{Sn} inside of its first radial node. The basis functions are then changed in the interstitial regions of the crystal only, where they should be, roughly speaking, made fit to meet Wigner-Seitz boundary conditions. Therefore it is expected (and confirmed by experience) that the optimum values of the variational parameters n_o and r_{Sn} are little affected by the crystal potential v and are basically determined by the geometry of the lattice. Hence, they may be obtained for any crystal structure once forever independently on the potential.

A striking feature of this approach is that for an LCAO calculation to be carried out there need not at all be a crystal potential present. It can be done for the empty lattice resulting in a well represented parabolic band structure and in plane waves well reproduced by their expansion (2.51) into a lattice sum of local functions. Just for this case the optimization procedure for the parameters n_o and r_{Sn} is carried out as described in the next section.

This consequent abandonment of the tight-binding point of view has two additional advantages practically very important. First of all, as the presence of a potential well v_S is no longer necessary for the construction of a basis function from (3.1), one is no longer forced to locate such functions only on atom sites. In many variants of LCAO, especially in that described in chapter 2, it is desirable to compose the crystal potential of spherical wells and to be able to restrict the basis to low angular momenta. This is possible for close-packed structures which typically metals crystallize into. For open semiconductor lattices, as e.g. the diamond lattice, on the one hand a superposition of spherical potential wells at the atom positions would be only a crude approximation for the crystal potential, and on the other hand it is known that even f-basis states are needed to get a good representation of conduction band states. The latter becomes immediately clear, if one realizes that a semiconductor conduction band corresponds to nearly-free electron states with rather large **k** and that in an angular momentum expansion of a plane wave the number of angular momenta needed rises with increasing **kr**. If we put additional potential sites in the interstitial centres of the diamond lattice, we obtain a b.c.c derivative two-component close-packed structure, where all distances to the next lattice point are reduced. The result is a considerably better potential representation by spherical wells (and hills) and a reduced number of angular momenta needed in the basis (due to reduced values of **kr**) [49].

The second point regards the connection of the optimum r_{Sn}-values with the lattice constants. Since the variational parameters r_{Sn} are initially not fixed, the only length parameters in an LCAO band structure calculation for the empty lattice are the lattice constants. Hence, the optimum r_{Sn}-values scale in a natural way with the lattice constants and are thus expressed by universal factors for any lattice type.

If we put $v_S=0$ in (3.1) and separate this equation in spherical coordinates we find, analogously to (2.7),

$$\left[-\frac{1}{r}\frac{d^2}{dr^2}r+\frac{l(l+1)}{r^2}+2\left(\frac{r}{r_{Sl}}\right)^{n_o}\right]\varphi_l(r)=\eta_{Sl}\,\varphi_l(r). \tag{3.2}$$

This equation is invariant under the scaling transformation

$$r \longrightarrow tr,\quad r_{Sl}\longrightarrow t^n r_{Sl},\quad \eta_{Sl}\longrightarrow t^{-2}\eta_{Sl},\quad n=\frac{n_o+2}{n_o}\,. \tag{3.2}$$

Therefore, if $r_{Sl}^{(a)}$ is the optimum value of r_{Sl} for a lattice with the lattice constant a, then

$$x_{Sl}=r_{Sl}^{(a)}\;/\;a^n=\text{inv}. \tag{3.4}$$

is an invariant. In other words the optimum r_{Sl}-values scale with the n-th power of the lattice constant and need be determined only for one value of the lattice constant.

3.2. Empty lattice tests

For a given lattice type, say f.c.c., the procedure of finding optimum parameters of (3.2) is as follows: fix the lattice parameter and choose values n_o and $r_{Sl}=r_l$ (in the case of a Bravais lattice as f.c.c. only $S=0$ appears). Solve (3.2) with the method described in section 2.1, take the lowest l-state for each $l=0,\ldots,l_{max}$ (l_{max} typically being 1 or 2) and perform a band structure calculation as described in the preceding chapter for a representative set of $\mathbf{k}$-points. Then vary the parameters and repeat the calculation. Typical results are shown in fig. 3.1.

Note first that all the computed energies stay above the exact values. If any energy is obtained below the expected one, this is usually due either to an insufficient number of two-centre integrals included or to a numerical instability ("overcompleteness") appearing at too large r_l-values and manifesting in a determinant of the overlap matrix $\mathbf{B}$ tending to zero. Next, a rapid increase of the energies for small r_l is seen the reason of which is completely obvious: the values approach the (increasing with decreasing r_l) η_l-values of (3.2). But not every function $\varepsilon_\nu(\mathbf{k};r_l)$ reaches a minimum when r_l increases. This may again easily be understood at least for the bottom of the band (lowest state at point Γ) the true wavefunction of which is a constant. Of course, the more extended the localized function in the lattice sum (2.50) (for $\mathbf{k}=0$) is the better a constant is approximated by this sum. Clearly, this behaviour is specific for the empty lattice. But even if one repeats the whole procedure with a crystal potential present [35], the minima of the $\varepsilon_\nu(\mathbf{k};r_l)$-curves are rather flat. Since the number of two-centre matrix elements which need be calculated increases with the $2n_o$-th power of r_l and the number of three-centre terms even more rapidly, the smallest value of r_l the deviation of $\varepsilon_\nu(\mathbf{k};r_l)$ from the true band energy does not exceed an admissible limit for should be chosen as the optimum.

In order to reduce the number of variational parameters, extended test calculations have been performed for the empty f.c.c. and b.c.c. lattices with various integer n_o-values with the issue that the best results are obtained for $n_o=4$. Furthermore, we use the atom radius r_{at} connected with the atom volume V_{at} in the lattice by

$$4\pi r_{at}^3/3 = V_{at} \tag{3.5}$$

instead of the lattice constant a as the reference length in (3.4), and fix once forever

$$\boxed{n_o = 4 \quad , \quad x_{Sl} = \frac{r_{Sl}}{r_{at}^{3/2}} = \text{inv.}} \tag{3.6}$$

The remaining parameters x_{Sl} (x_l in a lattice in which all atom sites are equivalent by symmetry) are conveniently obtained in the following manner. First vary x_o and x_1 and consider energy levels $\varepsilon_\nu(\mathbf{k})$ of pure s- and p-character, respectively. In a high symmetry structure there are always several low lying levels of such a type. This allows to find the optimum x_o and x_1 values independently (fig. 3.1a). Then, with these parameters fixed, x_2 is varied where now all levels may be considered (fig. 3.1b). The corresponding results for the b.c.c. and the h.c.p. lattices are shown on fig. 3.4 and fig. 3.6, respectively. All these curves were obtained by using 9 basis states per atom: one radial function for l=0,1,2 each. In the h.c.p. case the LCAO matrices are transformed into real ones with the help of the procedure described at the end of section 2.4. As is seen in the figures, some of the higher levels stabilize well above their true values. This indicates the lack of the next local states (2s and 4f) in the basis.

For all three basic metallic structures, f.c.c., b.c.c. and h.c.p.,

$$x_o = x_2 = 1.0 \quad , \quad x_1 = 0.95 \tag{3.7}$$

were found as optimum values. In fig. 3.2, 3.5, and 3.7 the free-electron band structures on symmetry lines as obtained by LCAO calculations using the basis parametrization (3.6-7) and 9 states per atom are shown. The cubic lattice constant was chosen $a=2\pi$ in atomic units which is in the range characteristic for metals and gives simple numbers for the true free electron band energies on symmetry points (the points X and H, respectively, correspond to $\mathbf{k}=(1,0,0)$ a.u.). Analogously, in the hexagonal lattice the lattice constants were chosen such that the M-point is $\mathbf{k}=(1,0,0)$ a.u. and the (c/a)-ratio is $\sqrt{8/3}$ corresponding to close-packed balls. The first LCAO band lies nearly constantly 1 mHartree (for b.c.c. and h.c.p.) to 2 mHartree (for f.c.c.) above the true one. In the second band the deviation near points of high symmetry is a little bit higher, up to 6 mHartree in the rather high unoccupied Γ-level, but the mean deviation is as previously. It starts rising only in the next higher bands. The loss of accuracy at symmetry points is typical for variational methods and corresponds to a symmetry restriction of the variational freedom: only definite combinations of the basis functions are allowed by symmetry. For instance, in the cubic structures and for the basis used, the coefficients $C_{n;\nu}(\mathbf{k})$ of (2.51) for $\mathbf{k}=0$ and, for certain bands ν, also at other $\mathbf{k}$-points of high symmetry are completely determined by symmetry. There is no variation besides that of n_o and x_1. This yields another justification for using just the energies at high symmetry $\mathbf{k}$-points in determining the optimum x_1-values.

Analogous LCAO free-electron bandstructures with only 4 basis states per atom (s and p) are shown on figures 3.3 and 3.8 for the f.c.c. and h.c.p. lattices, respectively (the b.c.c. case is comparable with f.c.c.). The results are basically unchanged for the first band, but become increasingly worse beginning with the top of the first band.

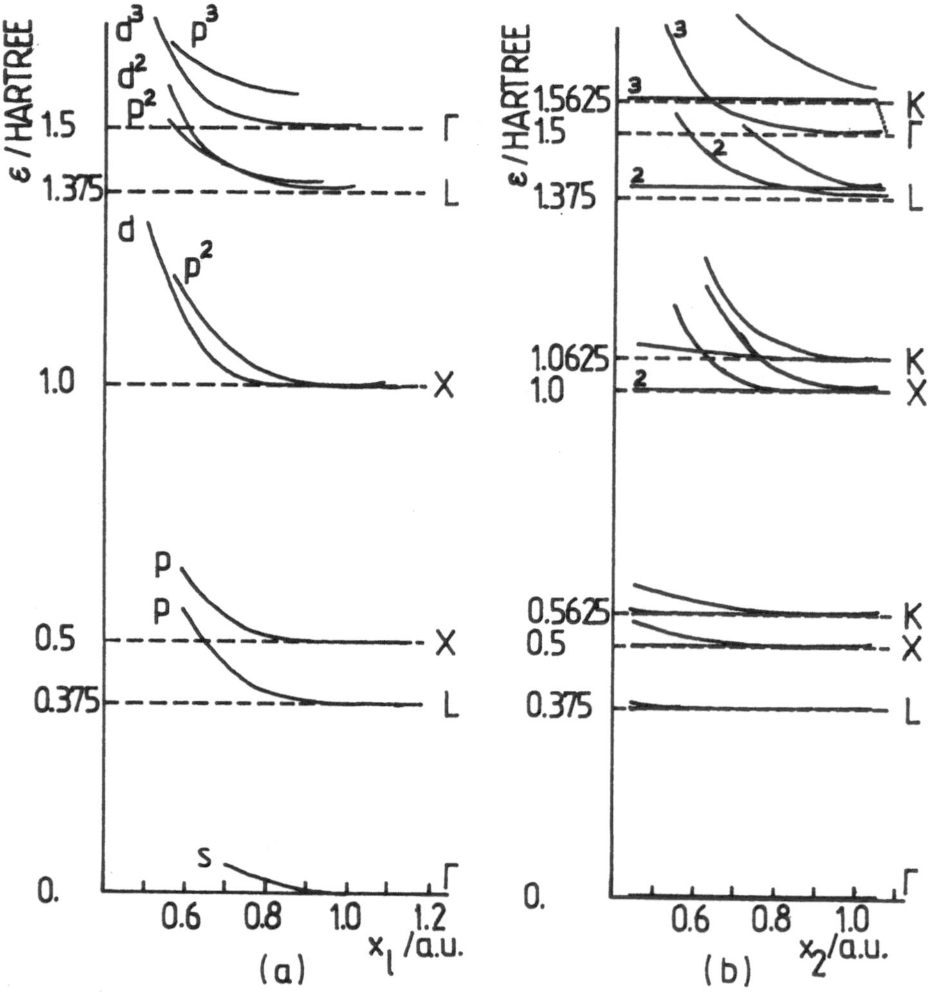

Fig. 3.1. Empty lattice tests for finding optimum parameters x_l of (3.6) for the f.c.c. lattice. LCAO band energies at the symmetry points Γ, X, L, and K of the B.Z. obtained from 9x9 LCAO matrices including atom-like s-, p-, and d-states constructed by means of equation (3.2) are plotted against the parameters x_l. The broken horizontals indicate the true free-electron energies for the lattice constant a=2π a.u. used in these calculations. In (a) the character of the wavefunction is labelled by small letters the exponent indicating multiplicity. In (b) only x_2 is varied, x_0=1. and x_1=0.95 being fixed at their optimum values found from (a). The small numbers at the curves again indicate the multiplicity. The incompleteness of the basis is first felt at the p^2-level at L staying well above the true value of 1.375 Hartree and then, more pronounced, at the p^3-level of 1.5 Hartree at Γ.

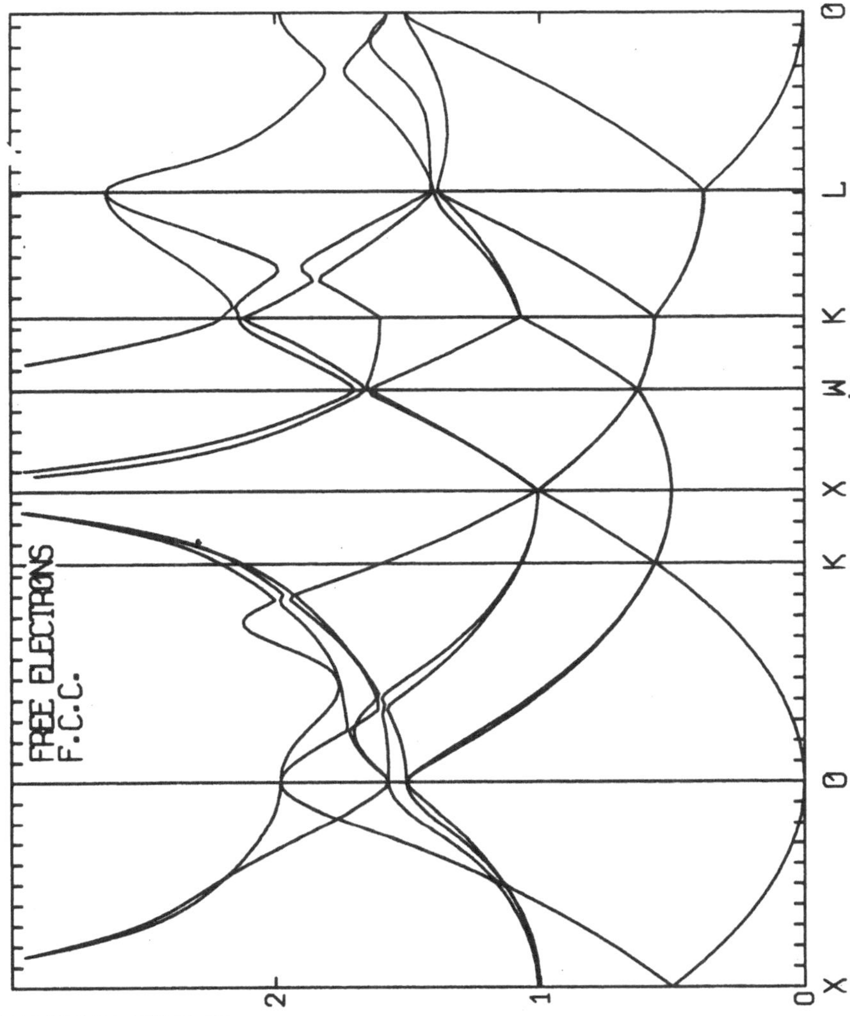

Fig. 3.2. The LCAO band structure for free electrons on symmetry lines (0 means the Γ-point) of the B.Z. of the f.c.c. structure obtained from 9x9 LCAO matrices including atom-like s-, p-, and d-states from equation (3.2) with optimum parameters (3.6-7). The energy values are for the cubic lattice constant a=2π a.u., but the results obtained with the same parameters (3.7) are the same for any lattice constant only the energies being rescaled.

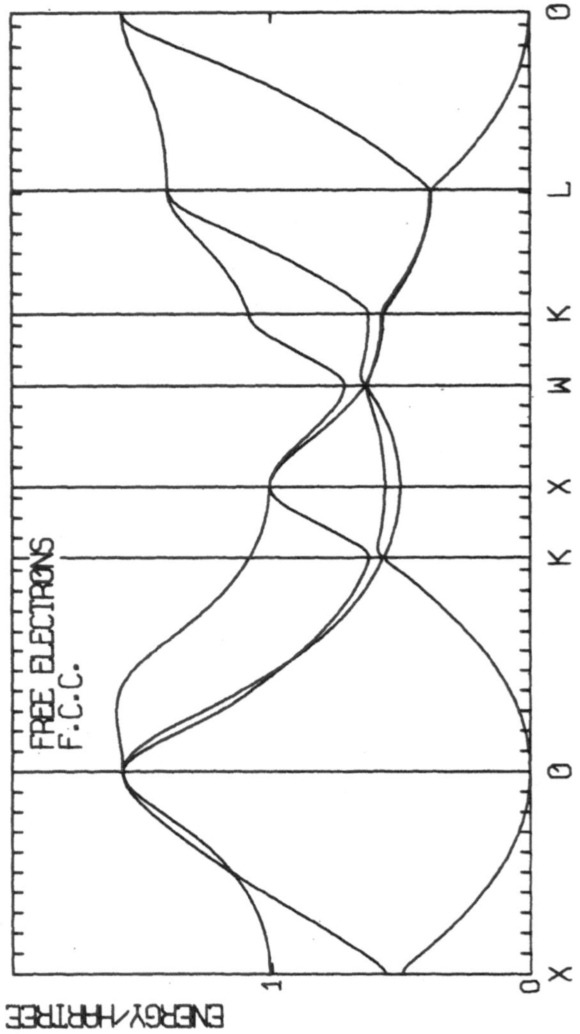

Fig. 3.3. The LCAO band structure for free electrons on symmetry lines of the B.Z. of the f.c.c. structure atom-like s- and p-states only. The results are otherwise obtained in the same manner as those of fig. 3.2. Noticable deviations due to the incompleteness of the basis first appear at point W being important for the Fermi surface of Al.

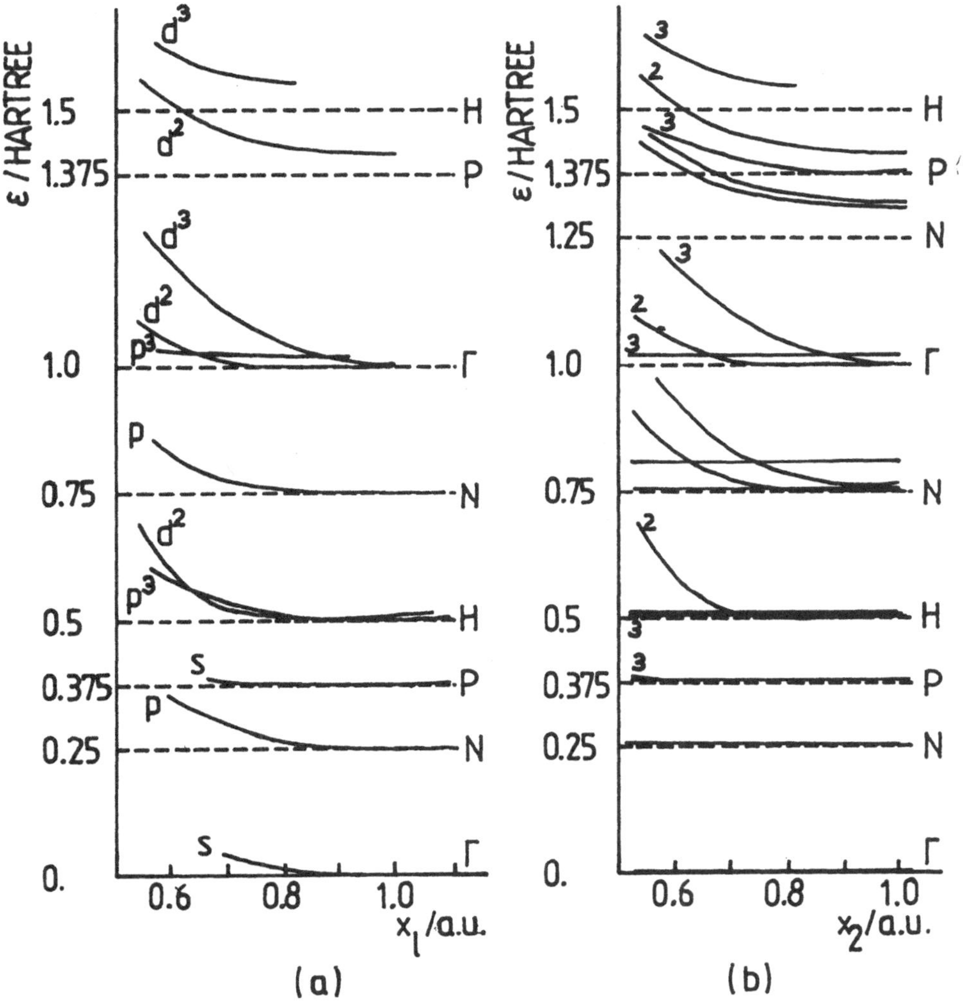

Fig. 3.4. Empty lattice tests for finding optimum parameters x_1 of (3.6) for the b.c.c. lattice. The representation is the same as in fig 3.1. The incompleteness of the basis is most drastically felt at the higher levels at point N. Besides the levels of pure s- and p-character shown in (a) there are some additional nearly horizontal lines in (b) corresponding to states of nearly s- and p-character, respectively.

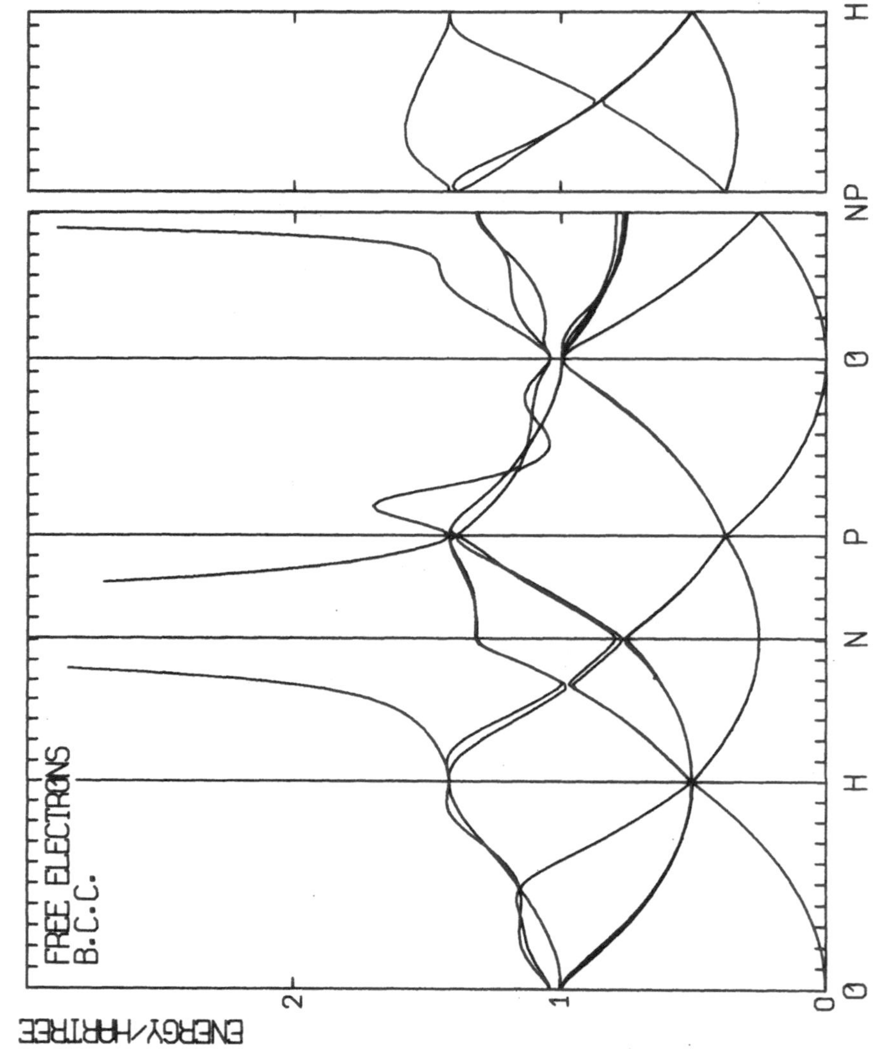

Fig. 3.5. The LCAO band structure for free electrons on symmetry lines of the B.Z. of the b.c.c. structure obtained from 9x9 LCAO matrices as in fig. 3.2. The bouncing up of the upper branch at point N is due to a numerical instability (det **B** tending to zero) appearing for the x_1-values (3.7) in the empty lattice case but disappearing as a crystal potential is switched on.

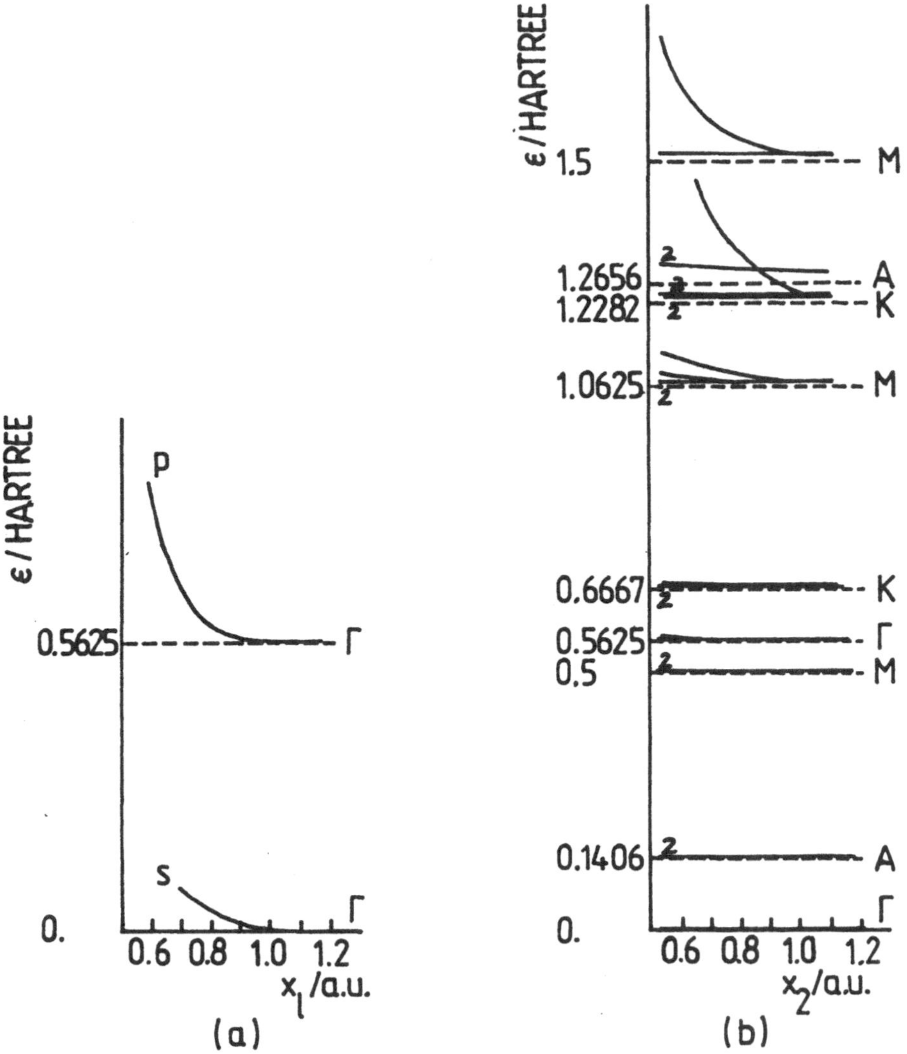

Fig. 3.6. Empty lattice tests for finding optimum parameters x_l of (3.6) for the h.c.p. lattice. The representation is the same as in fig. 3.1, but the LCAO matrices are now 18x18 matrices as there are two atoms per unit cell. The lattice constants were taken to be $a=2\pi/\sqrt{3}$ a.u. and $c=2\pi\sqrt{8/3}$ a.u., so that M corresponds to $\mathbf{k}=(1,0,0)$ a.u. All considered states below 1 Hartree, though mostly of mixed character, contain a very small d-admixture and appear therefore as nearly horizontals in (b). The incompleteness of the basis is practically not felt below the A-level at 1.2656 Hartree.

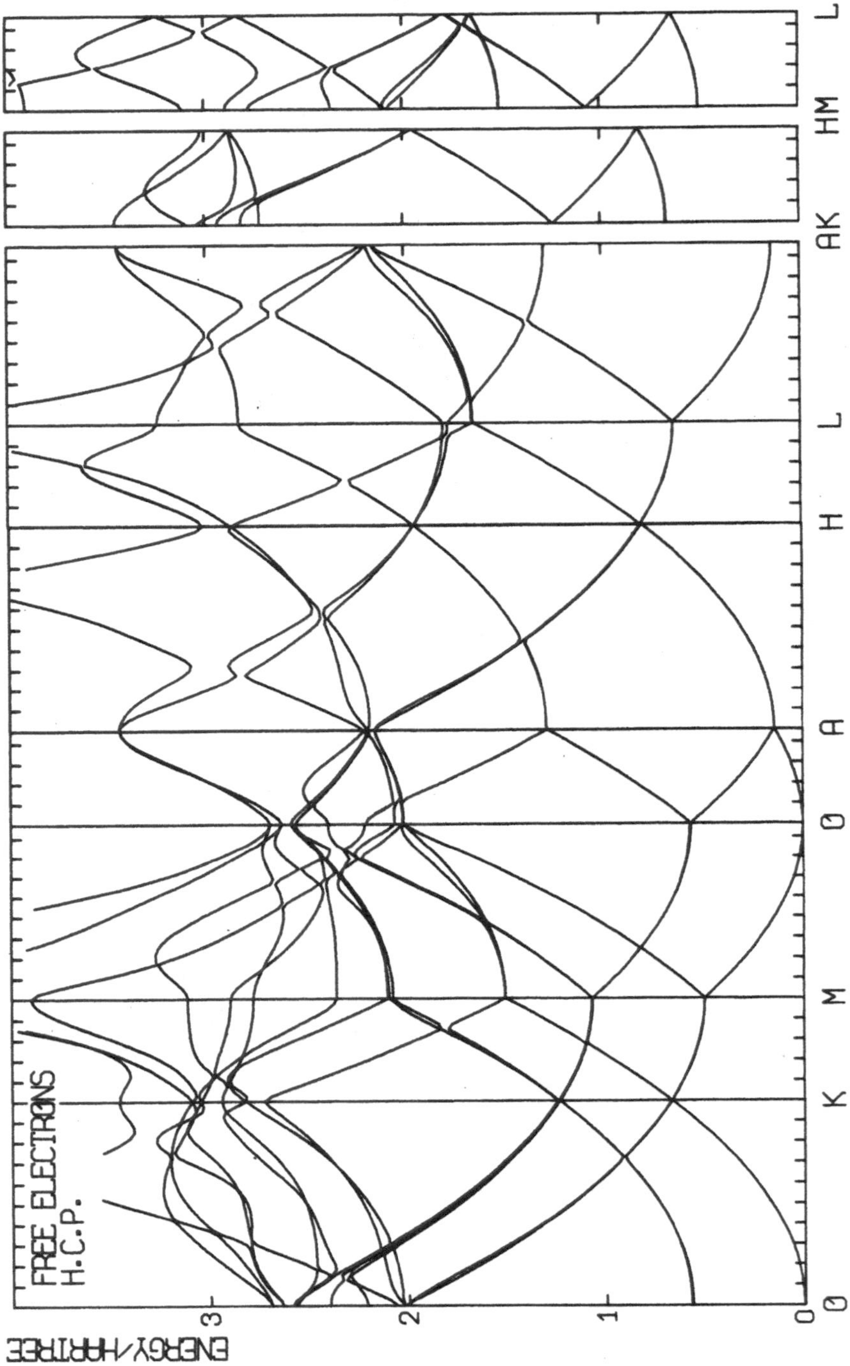

Fig. 3.7. The LCAO band structure for free electrons on symmetry lines of the B.Z. of the h.c.p. structure obtained from 18x18 LCAO matrices and otherwise as in fig. 3.2. The lattice constants are the same as in fig. 3.6.

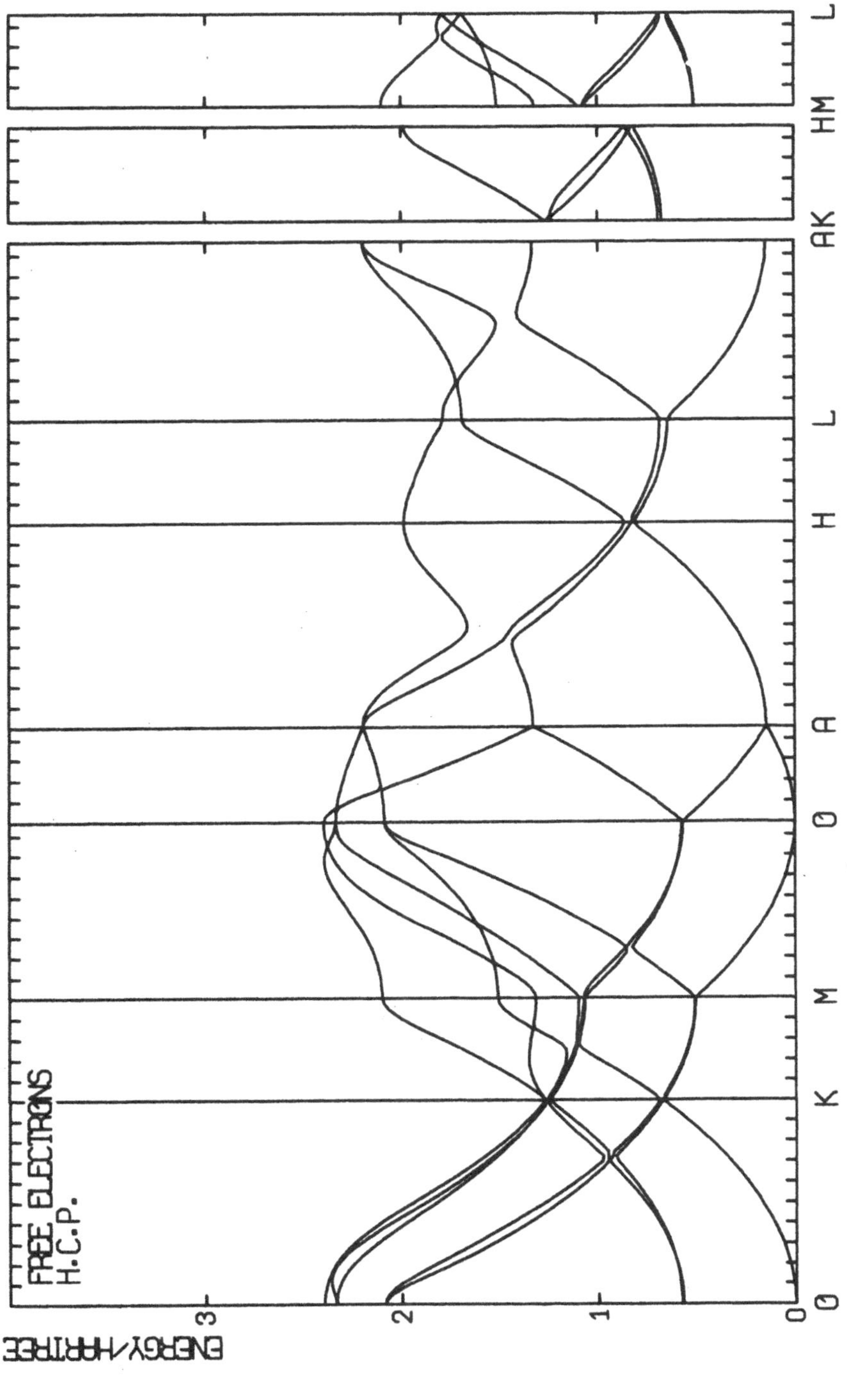

Fig. 3.8. The same as fig. 3.7, but obtained from 8x8 LCAO matrices including atom-like s- and p-states only.

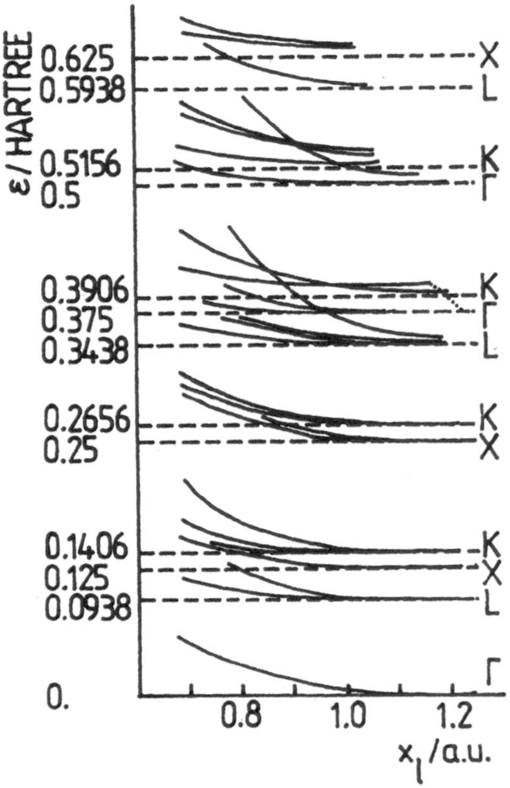

Fig. 3.9. Empty lattice test for finding optimum parameters x_l of (3.6) for the diamond lattice without basis states on interstitial sites and obtained from 18x18 LCAO matrices including atom-like s-, p-, and d-states. The lattice constant of the f.c.c. sublattice was chosen to be $a=4\pi$ a.u. being a bit larger than in the Si lattice. The plot is against $x=x_0=x_2=x_1/0.95$. All occupied levels of a group IV material up to the Γ-level at 0.375 Hartree are correctly obtained, but the first conduction state at that level (marked by a dotted line) strongly feels the lack of f-admixture. If basis states on interstitial sites are introduced, a b.c.c. lattice with a half as large lattice constant is obtained. The Brillouin zone of the diamond structure is now obtained by folding the b.c.c. Brillouin zone so that the points P and H coincide with Γ, and N coincides with X. The levels drawn here are then to be compared to those of fig. 3.4 at the same energies. There is no longer any incompleteness of the basis felt around the energy corresponding to the semiconductor gap even if one would omit the d-states.

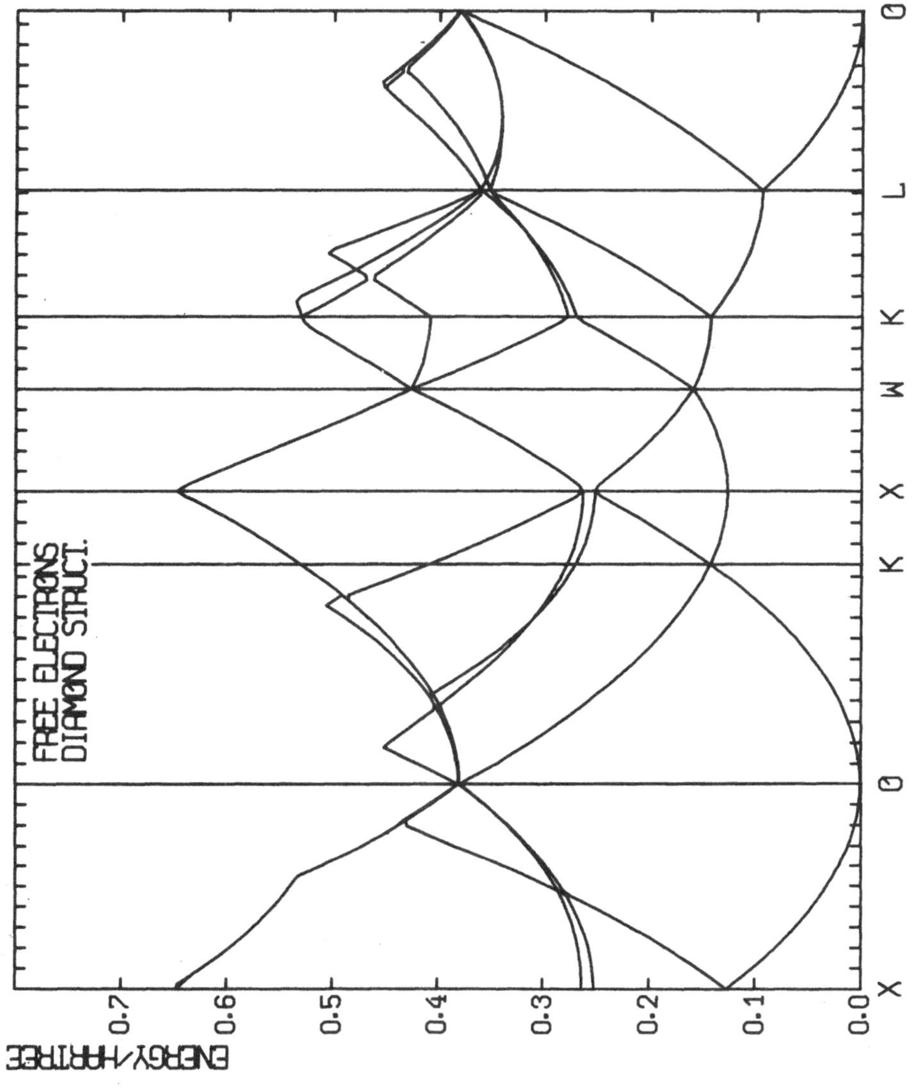

Fig. 3.10. The LCAO band structure for free electrons on symmetry lines of the B.Z. of the diamond structure obtained from 16x16 LCAO matrices including atom-like s- and p-states only located at atom and interstitial sites (4 sites per unit cell). Though the true band structure coincides with the f.c.c. case, the results shown are different from those of fig. 3.3 not only by a rescaling of energies (due to the doubled lattice constant): The basis states are located on different sites in real space.

As an example of an open structure, in figures 3.9 and 3.10 the diamond lattice is treated without and with interstitial states in the basis. These results are to be compared to the b.c.c. case, since the diamond lattice including interstitials sites (empty centres of tetrahedrons) is a b.c.c. derivative structure. The comparison illustrates what was discussed in the previous section.

3.3. Optimum basis LCAO band calculation for real crystals

We close this chapter with an illustration of how the optimum basis LCAO method works for the band structure calculation of real crystals. As a further illustration the results of self-consistent band structure calculations given in Appendix 3 may be considered.

As was already explained in section 3.1., the optimal r_{Sl}-values are little affected by the crystal potential. Therefore, the values of (3.4-6) are used in constructing the LCAO basis via (3.1) for any material crystallizing in the f.c.c., b.c.c. or h.c.p. structure. For any other structure just the empty-lattice test of last section is to be repeated resulting in universal (with respect to volume and potential changes) parameters (3.4) for that structure type.

The first question that arises is how many basis states should be used in order to get reliable results for the valence and conduction bands. This question can be generally answered at least for any metallic material. (Up to now we have little experience with semiconductors, but the general reasoning should be just the same.) Metals are commonly divided into two classes: "simple" metals having only nearly-free electron conduction bands (but being generally by no means simple) and transition metals having in the crystalline state roughly one nearly-free conduction electron per atom and besides resonant (tight-binding) d- and/or f-states in the conduction band region. Two electrons per unit cell go in each band. Consequently, in accordance with the results of the last section, only s- and p-nearly-free electron basis states are needed for the alkaline and eventually the earth-alkaline metals and for all transition metals. For the higher valent metals the nearly-free d-electron basis states must be included being e.g. necessary to obtain the correct Fermi surface of Al near point W (cf. fig. 3.2 and 3.3 as well as the Al band structure of Appendix 3). For transition metals, tight-binding d- and/or f-states must be added to the basis. Though being of tight-binding character, yet the corresponding states of the free atom have long-ranged tails [116] and should therefore be compressed using equation (3.1). As is well known the d-resonance being above the occupied states of the heavier earth-alkaline metals Ca, Sr, Ba and Ra is strongly felt by the electrons on the Fermi level [114]. Conversely. the core d-states of Zn, Cd and Hg hybridize with the valence states at the band bottom. Hence, d-states should be included in the LCAO basis also in these cases.

Summarizing it can be stated that for any metal atom just one basis state per angular dependence is needed, s- and p-states only for the alkaline metals and Be and Mg, and s-, p- and d-states in all other cases except for the lanthanides and actinides where also f-states must be included. (Of course a relativistic LCAO version [106] is needed for the heavier elements and one must be aware of more complicated correlation effects for f-electrons.)

Note the very important fact that we need not at all worry about the completeness of the LCAO basis in the inner range of the atoms because this is always safeguarded by the special properties of atom-like functions discussed in section 2.1.

For illustration we show on fig. 3.11 the results of an optimum basis LCAO band calculation for Cu using Chodorow's potential (expanded according to (2.20-2.25)) in comparison with the APW results of [16]. The basis (2.10) was used for the expansion of radial functions with the only difference that the smallest α-value was related to the atom radius by

$$\boxed{\alpha_{min} = 4.4/r_{at}} \tag{3.8}$$

in atomic units. This relation was also used in all calculations the results of which are given in Appendix 3. The actual radial basis in the copper calculation was

$$b_{l\nu\alpha}(r) = r^{l+\nu-1}\, e^{-\alpha r},$$
$$\nu = 1,2,3\ ;\ \alpha = 29.,\ 11.01,\ 4.18,\ 1.59.$$

The convergency of the lattice sums of (2.53) and (2.55) was checked and four neighbours were found to suffice for a 1 mHartree accuracy. The three-centre terms (last line of (2.64)) were included only for configurations where both wavefunction centres are first neighbours of the pseudopotential centre. The deviation of the band energies from Burdick's APW results were generally less than 10 mHartree (which is of the order of the accuracy of Burdick's calculation) exept for the top of the p-bands where the incompleteness of the basis is felt. The deviations are partly caused by the expansion of the muffin-tin potential having a singularity at the muffin-tin radius (outside of which it is constant [16]) into a Slater function basis.

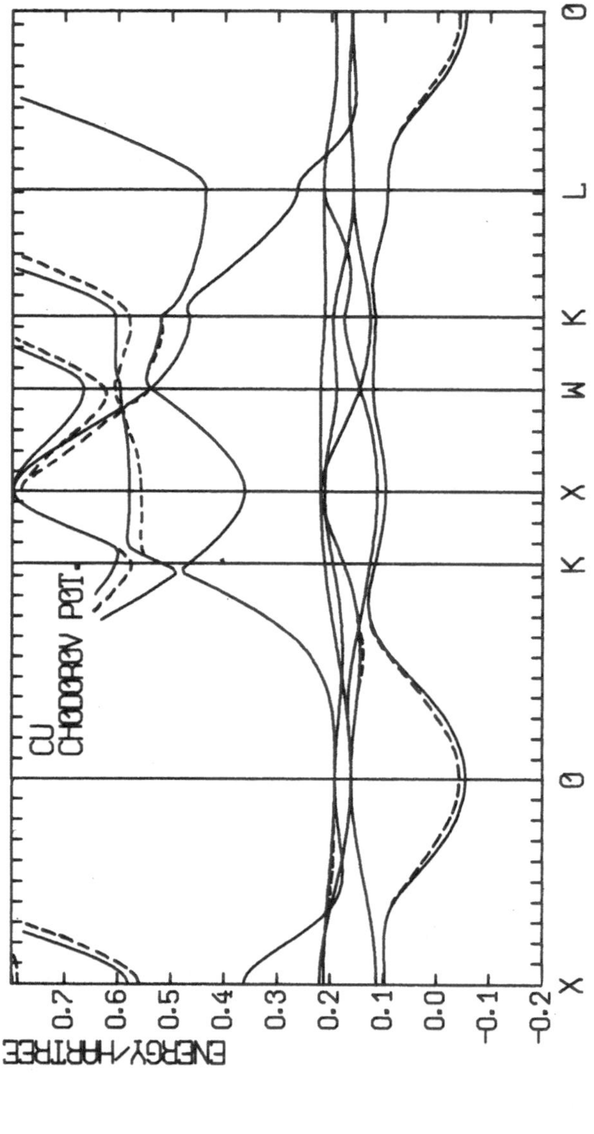

Fig. 3.11. LCAO band structure of Cu (full curves) obtained by using Chodorov's potential and compared to the APW results of [16] (broken curves). The LCAO treatment was a refined version of [35].

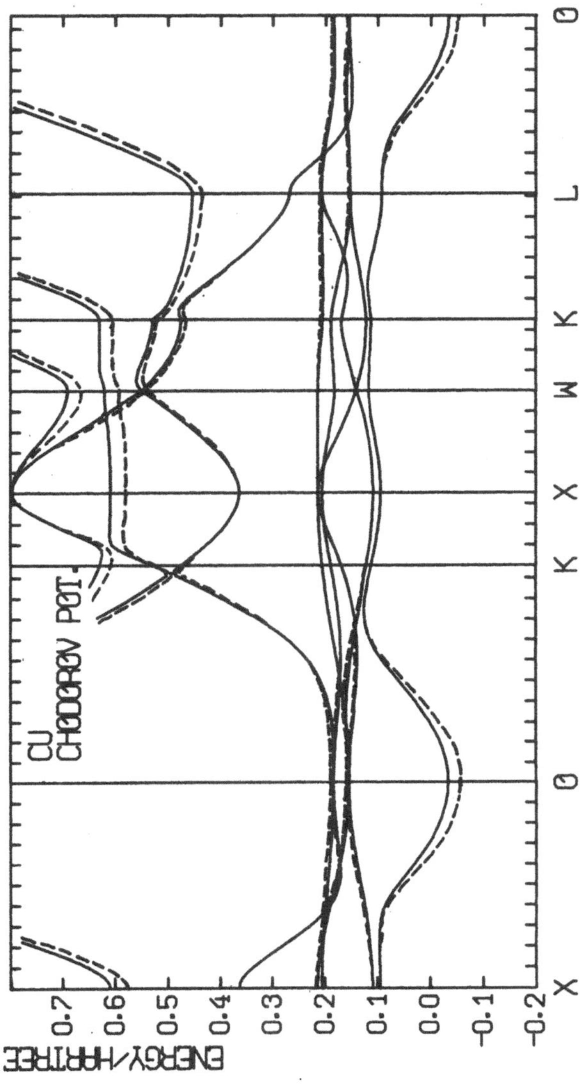

Fig. 3.12. The LCAO bandstructure of Cu as in fig. 3.11 (broken curves). The full curves correspond to results with all three-centre integrals and crystal field terms neglected. The changes are generally less than 20 mHartree in the occupied bands.

In fig. 3.12 the results of a repeated calculation are shown where all three-centre pseudo-potential matrix elements and all crystal field terms are neglected. The changes are rather small due to the weakness of the pseudo-potential. These results are in remarkable contrast to the heavy convergency problems of early LCAO applications to metals [81]. The smallness of the total contribution of three-centre terms justifies the general neglect of the $m_a \neq m_b$ contributions in the geometry of fig. 2.4 simplifying the Slater-Koster parametrization of the matrix elements considered in section 4.1.

4. WANNIER FUNCTIONS

This central chapter, placed between the first part of this volume concerned with the problems of a bandstructure calculation for a given crystal potential and the second part concerned with the construction of the self-consistent potential, plays the role of a caesura. It gives a survey over the various representation theoretical problems with local bases being important for many applications. The presentation begins with an analysis of LCAO interpolation schemes in comparison with full wave-function based LCAO calculations. In the following two sections the Wannier representation, implicitly underlying all LCAO interpolation schemes and a great variety of model Hamiltonians, is introduced and analyzed. The variation principle for Wannier functions is derived and used to find a direct link of our optimum atom-like wavefunction representation to a certain Wannier representation. It is illustrated in section 4.4. on the example of the copper band structure. An important application of local basis representations benefitting by the ortho-normality relation is the recursion method for the electronic structure of finite or disordered systems. The chapter closes with a short discussion of Anderson's chemical pseudopotential closely linked to the previous problems.

4.1. LCAO interpolation

The band structures shown on the figures of chapter 3 and of Appendix 3 may despite of their complexity be characterized by a few numbers: the symmetry reduced matrix elements contained in the lattice sums (2.53) and (2.55). This was first pointed out by Slater and Koster [113]. In the extreme tight-binding case, where the **B** matrix (2.53) may be replaced by the unit matrix, these parameters are indeed more or less uniquely determined to be the Hamiltonian matrix elements formed with tightly bound atomic wavefunctions or, from another point of view, to be the Fourier components of the **k**-dependend matrix elements $H_{\mathbf{S}'m;\mathbf{S}n}(\mathbf{k})$ of (2.55) in an angular momentum representation.[1)]

---

1) To be more precise, the representation should use lattice harmonics corresponding to irreducible representations of the point groups of the lattice sites instead of spherical harmonics. But at least for the octahedron point group appearing in cubic lattices the real spherical harmonics up to l=2 coincide with the lattice harmonics, and in many other cases the additional matrix elements allowed for in a lattice

(cont. on p. 74)

This latter point is used in LCAO interpolation schemes: Make an ansatz

$$H_{\mathbf{S}'m;\mathbf{S}n}(\mathbf{k}) =$$

$$= \sum_{\mathbf{R}} e^{i\mathbf{k}(\mathbf{R}+\mathbf{S}-\mathbf{S}')} \left[\sum_{pq} a_{mp}(\mathbf{R}+\mathbf{S}-\mathbf{S}')\ \varepsilon_{pq}(\mathbf{R}+\mathbf{S}-\mathbf{S}')\ a^{+}_{nq}(\mathbf{R}+\mathbf{S}-\mathbf{S}') \right] \tag{4.1}$$

where the $\varepsilon_{pq}(\mathbf{R}+\mathbf{S}-\mathbf{S}')$ are the matrix elements remaining independend after the symmetry reduction, and $\mathbf{a}(\mathbf{R}+\mathbf{S}-\mathbf{S}')$ are the transformation matrices of the symmetry reduction (by nature transformations from local to global coordinates). Get the matrix elements on the left hand side of (4.1) for a representative set of $\mathbf{k}$-values from anywhere, and fit the parameters $\varepsilon_{pq}(\mathbf{R}+\mathbf{S}-\mathbf{S}')$ to these figures. In practice usually the band energies being the eigenvalues of the matrices $\mathbf{H}$ of (4.1) are used instead of the matrix elements in the fit. (Remember that the $\mathbf{B}$-matrix of (2.53) had been replaced by the unit matrix.)

So far for the tight-binding case. But of course, being apart from the symmetry reduction just a Fourier expansion, (4.1) may be used as a Hamiltonian matrix in some (not explicitely given) orthonormal local basis representation the existence of which will be shown in section 4.3. for <u>any</u> band structure. It will however be seen there that with the local basis being no longer fixed from the outset a vast ambiquity comes in the game, and both skill and experience are needed to make a convenient and physically suggestive choice [57]. It must be emphasized in this connection that for a complex band structure all the complexity is now cast into the orthonormal local basis states which may turn out to be rather exotic [1,2] and to have little in common with atom-like states. The use of the representation (4.1) for the band structure in combination with matrix elements of atom-like states in the calculation of observable characteristics may be completely misleading. It is the main drawback of LCAO interpolation schemes that they do not provide any direct information about wavefunctions.

Of course, to avoid complicated orthonormal local basis states a simultaneous fit of the matrices $\mathbf{H}(\mathbf{k})$ and $\mathbf{B}(\mathbf{k})$ of (2.55) and (2.53) could be tried, but this would even increase the ambiguity. In order to illuminate the situation we give in table 4.1 the values of the Slater-Koster parameters of the $\mathbf{H}$ and $\mathbf{B}$ matrices corresponding to our optimum

---

(cont. from p. 73)

harmonics representation may be neglected as discussed in section 2.4 in connection with the three-centre integrals. As we shall not use it we do not introduce the business of lattice harmonics here.

Table 4.1. The LCAO matrix elements of the band structure of the empty f.c.c. lattice shown in fig. 3.2.

Hamiltonian (in Hartree):

one-centre:	(s)	(p)	(d)
	0.2697	0.5652	0.7702

two-centre, 1. to 8. nearest neighbour:

$(ss\sigma)$	$(ps\sigma)$	$(ds\sigma)$	$(pp\sigma)$	$(pp\pi)$	$(dp\sigma)$	$(dp\pi)$	$(dd\sigma)$	$(dd\pi)$	$(dd\delta)$
-0.0086	-0.0254	-0.0889	-0.1339	0.0057	-0.2352	0.0674	0.2258	-0.2250	0.0330
-0.0100	0.0146	0.0086	0.0191	-0.0052	-0.0103	-0.0083	0.0759	-0.0051	-0.0028
-0.0033	0.0055	0.0080	0.0109	-0.0012	0.0148	-0.0037	-0.0088	0.0085	-0.0011
-0.0010	0.0015	0.0027	0.0028	-0.0002	0.0059	-0.0008	-0.0098	0.0030	-0.0002
-0.0003	0.0003	0.0007	0.0005	0.0	0.0014	-0.0001	-0.0036	0.0006	0.0
-0.0001	0.0001	0.0001	0.0	0.0	0.0002	0.0	-0.0009	0.0	0.0
0.0	0.0	0.0	0.0	0.0	-0.0001	0.0	0.0	-0.0001	0.0
0.0	0.0	0.0	0.0	0.0	-0.0001	0.0	0.0001	0.0	0.0

overlap:

one-centre:	(s)	(p)	(d)
	1.0	1.0	1.0

two-centre, 1. to 8. nearest neighbour:

$(ss\sigma)$	$(ps\sigma)$	$(ds\sigma)$	$(pp\sigma)$	$(pp\pi)$	$(dp\sigma)$	$(dp\pi)$	$(dd\sigma)$	$(dd\pi)$	$(dd\delta)$
0.1728	-0.2527	-0.2825	-0.3758	0.1044	-0.3339	0.2313	0.1745	-0.3862	0.1114
0.0323	-0.0483	-0.0743	-0.0871	0.0100	-0.1322	0.0327	0.1930	-0.0969	0.0110
0.0069	-0.0084	-0.0138	-0.0126	0.0006	-0.0283	0.0028	0.0589	-0.0141	0.0005
0.0016	-0.0013	-0.0016	-0.0006	-0.0001	-0.0025	-0.0004	0.0103	-0.0002	-0.0002
0.0004	-0.0001	0.0003	0.0005	-0.0001	0.0011	-0.0004	-0.0004	0.0010	-0.0001
0.0001	0.0	0.0004	0.0003	0.0	0.0009	-0.0002	-0.0015	0.0006	-0.0001
0.0	0.0	0.0002	0.0002	0.0	0.0005	-0.0001	-0.0010	0.0002	0.0
0.0	0.0	0.0001	0.0001	0.0	0.0002	0.0	-0.0005	0.0001	0.0

atom-like basis of chapter 3. and leading to the free-electron band-structure of fig. 3.2 for the empty f.c.c. lattice. The point symmetry group of the Hamiltonian of an empty lattice at any lattice site is the group O(3) of all orthogonal transformations in three dimensions, and therefore the angular momentum representation is here the correct one. The z-axis of the local coordinate system for the matrix elements $\varepsilon_{pq}(\mathbf{R}+\mathbf{S}-\mathbf{S}')$ is chosen to be the direction of $\mathbf{R}+\mathbf{S}-\mathbf{S}'$, and $\mathbf{a}(\mathbf{R}+\mathbf{S}-\mathbf{S}')$ are the transformations rotating the spherical harmonics and given in Appendix 1. The two-centre matrix elements $\varepsilon_{pq}(\mathbf{R}+\mathbf{S}-\mathbf{S}')$ are denoted in the usual manner by $(l_p l_q m)$, e.g. $(dp\sigma)$ meaning $l_p=2$, $l_q=1$ and $m=m_p=m_q=0$. For instance, for the matrix element $H_{ss}(\mathbf{k}=\mathbf{0})$ decoupled by symmetry we find from table 4.1 and with the coordination numbers of the f.c.c. lattice

$$H_{ss}(\mathbf{0}) = 0.2697 - 12\cdot 0.0086 - 6\cdot 0.0100 - 24\cdot 0.0033 - \\ - 12\cdot 0.0010 - 24\cdot 0.0003 - 8\cdot 0.0001 = 0.0073,$$

and analogously

$$B_{ss}(\mathbf{0}) = 3.4626$$

resulting in the energy

$$\varepsilon_s(\mathbf{0}) = H_{ss}(\mathbf{0}) / B_{ss}(\mathbf{0}) = 0.0021$$

of the band bottom. At point X we find

$$H_{ss}((1,0,0)) = 0.2637 + 4\cdot 0.0086 - 6\cdot 0.0100 - 8\cdot 0.0033 -$$
$$- 12\cdot 0.0010 + 8\cdot 0.0003 - 8\cdot 0.0001 = 0.2073,$$
$$B_{ss}((1,0,0)) = 0.5746,$$
$$\varepsilon_s((1,0,0)) = 0.3608.$$

Neglecting the dispersion of the **B**-matrix would reduce the bandwidth almost by a factor of two (fig. 4.1). This situation develops dramatically in the p-band where the matrix element $B_{pp}(\mathbf{0})=0.1058$ pushs up the band energy by a factor of about ten to its value of about 1.5 (third level at point $\Gamma\equiv 0$ in fig. 3.2).

Even more important than merely broadening the band is the role of the dispersion of $\mathbf{B}(\mathbf{k})$ in producing a strong curvature at the top of the bands: A Fourier component of $\mathbf{B}(\mathbf{k})$ of the order of one going in the denominator of the energy expression produces an infinite series of Fourier components in the numerator according to

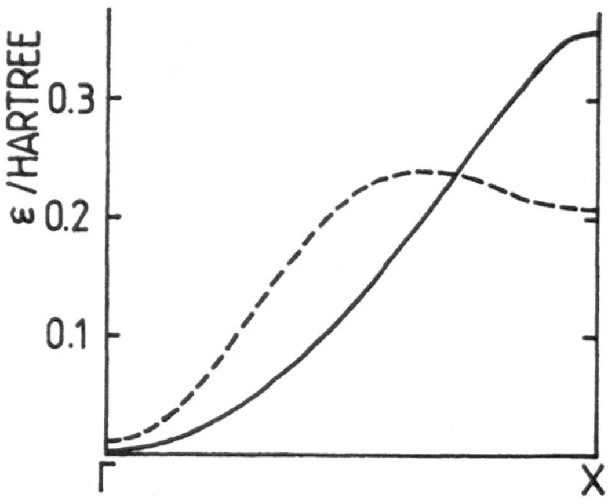

Fig. 4.1. An unhybridized LCAO-s-band on the line Γ-X for the empty f.c.c. structure is shown with (full curve) and without (broken curve) taking into account the overlap matrix element $B_{ss}(\mathbf{k})$. The latter results in a band broadening and especially in a narrowing of the top of the band according to (4.2) thus increasing the curvature.

$$(1 - ae^{ikR})^{-1} = \sum_{n=0}^{\infty} a^n e^{iknR} , \tag{4.2}$$

just these high Fourier components being indispensable for a strong curvature of a band like the upper s-p-band on the line Γ-X of the f.c.c. structure near the Γ-point. (See fig. 3.2. The even stronger curvature of the other bands nearby is a result of hybridization, and no high Fourier component is needed in these cases; cf. the situation with one-band and multi-band Wannier functions below.)

These considerations suggest another LCAO interpolation scheme for nearly-free electron bands which has not yet been tried up to now: Take the Hamiltonian matrix as a constant times the unit matrix and try an LCAO interpolation of the **B**-matrix only. In this case, on the one hand a smaller number of parameters should generally suffice, and on the other hand the interpolation might on a safer base be considered as corresponding to atom-like local basis states with large overlap integrals. It would essentially mean an LCAO interpolation scheme for the band structure $\varepsilon^{-1}(\mathbf{k})$ of the operator $\hat{h}^{-1}$, and the existence of the corresponding representation is proved exactly along the same lines as for the Wannier representation in the subsequent sections.

4.2. One-band Wannier functions

Consider the following isometric transformation of the Bloch states $|\mathbf{k}\nu\rangle$ of the ν-th band into Wannier states denoted in the following by curled brackets:

$$|\mathbf{R}\nu\} = \frac{1}{\sqrt{N}} \sum_{\mathbf{k}} |\mathbf{k}\nu\rangle\, e^{-i\mathbf{k}\mathbf{R}-i\alpha_\nu(\mathbf{k})} . \tag{4.3}$$

The **k**-summation was defined in (2.49). As a solution of the homogeneous linear equation (2.3) the state $|\mathbf{k}\nu\rangle$ is determined up to a constant phase factor $\exp(-i\alpha_\nu(\mathbf{k}))$ having been explicitly introduced in (4.3). In full notation (4.3) means

$$\{\mathbf{r}|\mathbf{R}\nu\} = \{\mathbf{r}-\mathbf{R}|0\nu\} \equiv \varphi_\nu(\mathbf{r}-\mathbf{R}) =$$

$$= \sqrt{N}\, \frac{V_{U.C.}}{(2\pi)^3} \int_{B.Z.} d^3k\, \varphi_{\mathbf{k}\nu}(\mathbf{r})\, e^{-\mathbf{k}\mathbf{R}-i\alpha_\nu(\mathbf{k})} , \tag{4.4}$$

The inverse transformation is

$$|\mathbf{k}\nu\rangle = e^{i\alpha_\nu(\mathbf{k})} \frac{1}{\sqrt{N}} \sum_{\mathbf{R}} |\mathbf{R}\nu\} \, e^{i\mathbf{kR}} \tag{4.5}$$

which is easily demonstrated by inserting (4.5) into (4.3). In the same way the equivalence of the orthonormalization relations

$$\langle \mathbf{k}'\mu|\mathbf{k}\nu\rangle = \delta_{\mu\nu}\,\delta_{\mathbf{k}'\mathbf{k}} \xrightarrow[N\to\infty]{} \delta_{\mu\nu} \frac{(2\pi)^3}{NV_{U.C.}} \delta(\mathbf{k}'-\mathbf{k}) \tag{4.6}$$

and

$$\{\mathbf{R}'\mu|\mathbf{R}\nu\} = \delta_{\mu\nu}\,\delta_{\mathbf{RR}'} \tag{4.7}$$

is shown demonstrating the isometry of the transformations (4.3) and (4.5). These considerations are at least correct as long as the crystal consists of a <u>finite</u> number N of unit cells (with periodic boundary conditions understood, the Wannier function being then periodic with the crystal volume as one period). The difficulties the limiting process $N \to \infty$ may cause are discussed below.

For the moment we realize that (4.5) is a Fourier expansion of the periodic function (with respect to **k**) $|\mathbf{k}\nu\rangle$ and introduce another Fourier expansion for $\varepsilon_\nu(\mathbf{k})$:

$$\varepsilon_\nu(\mathbf{k}) = \sum_{\mathbf{R}} e^{-i\mathbf{kR}}\,\varepsilon_{\mathbf{R}\nu}\,, \quad \varepsilon_{\mathbf{R}\nu} = \frac{1}{N} \sum_{\mathbf{k}} e^{i\mathbf{k}R}\,\varepsilon_\nu(\mathbf{k})\,. \tag{4.8}$$

If we now operate onto (4.3) for **R**=0 with the crystal Hamiltonian and use (4.8) we end up with the analogue of Schrödinger's equation for Wannier states,

$$\hat{h}|0\nu\} = \sum_{\mathbf{R}} |\mathbf{R}\nu\}\varepsilon_{\mathbf{R}\nu}\,, \tag{4.9}$$

being a difference differential equation in **r**-representation:

$$\left[-\frac{\Delta}{2} + v(\mathbf{r})\right]\varphi_\nu(\mathbf{r}) = \sum_{\mathbf{R}} \varphi_\nu(\mathbf{r}-R)\,\varepsilon_{\mathbf{R}\nu}\,, \tag{4.10}$$

if the $\varepsilon_{\mathbf{R}\nu}$ are known. A projection of (4.9) onto $|\mathbf{R}\mu\}$ and use of (4.7) gives

$$\{R\mu|\hat{h}|0\nu\} = \delta_{\mu\nu}\,\varepsilon_{R\nu}\ , \tag{4.11}$$

a very useful relation for the study of the asymptotics of Wannier functions (see below).

Some comments are in due place here on the vast ambiguity in the definition of Wannier states by (4.3). Not only the phase factor $\exp(-i\alpha_\nu(\mathbf{k}))$ may be chosen completely arbitrarily (particularly it needs not be a continous function of $\mathbf{k}$) also the set of states $|\mathbf{k}\nu\rangle$ forming a band is not given by itself. The numbering of states with a given $\mathbf{k}$ by the band index ν may again be chosen independendly for any $\mathbf{k}$. Of course, suggestive is a consecutive numbering with increasing energy for every $\mathbf{k}$ leaving undetermined only the cases of degeneracy, yet there are cases suggesting another approach. If for instance a broad nearly-free electron band as b-b of fig. 4.2 hybridizes with a flat tight-binding band (a-a in fig. 4.2), then it might be desirable to connect a certain Wannier function everywhere with the tight-binding band which enforces a discontinous definition in the hybridization area whereas the numbering with increasing energies always leads to continous functions $\varepsilon_\nu(\mathbf{k})$ (having discontinous derivatives in points of degeneracy). In particular, Wannier functions need not have any symmetry and they obey no superposition principle: linear combinations of differently defined Wannier functions, even for the same band, are generally not again Wannier functions.

In order to get some feeling of the properties of Wannier functions, consider first the formal similarity between the expressions (4.5), for $\alpha_\nu(\mathbf{k})\equiv 0$, and (2.50). The main difference is that the states on the right hand side of (2.50) are not orthonormal whereas the Wannier states are. However, in the extreme tight-binding case the overlap integrals of neighbouring atom-like states may be neglected so that atom-like states are good approximants of Wannier states of tight-binding bands, eventually after a first order orthogonalization by Löwdin's procedure [92].

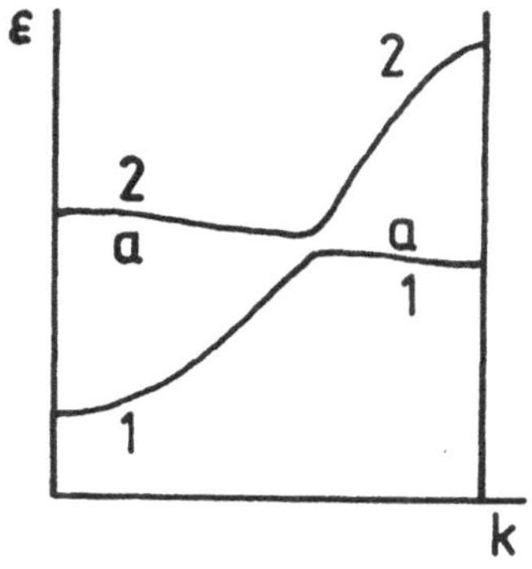

Fig. 4.2. Different possibilities of band numbering in a hybridized band structure.

Consider now the other extreme of a free electron band

$$\langle \mathbf{r} | \mathbf{k}1 \rangle = \frac{1}{\sqrt{NV_{U.C.}}} e^{i\mathbf{kr}} , \quad \mathbf{k} \in \mathrm{B.Z.} \tag{4.12}$$

of a close-packed Bravais lattice (f.c.c. say). We put again $\alpha_1(\mathbf{k}) \equiv 0$ and have

$$\{\mathbf{r} | 0,1\} = \varphi_1(\mathbf{r}) = \frac{\sqrt{V_{U.C.}}}{(2\pi)^3} \int_{\mathrm{B.Z.}} d^3k \; e^{i\mathbf{kr}} . \tag{4.13}$$

To come up with a qualitative statement we approximate the Brillouin zone by a sphere of Radius G/2 (G being of the order of the first reciprocal lattice vector) and find readily

$$\varphi_1(\mathbf{r}) \quad \frac{\sqrt{V_{U.C.}}}{8\pi^2} \frac{G^2}{r} j_1(Gr/2) =$$

$$= \frac{\sqrt{V_{U.C.}}}{4\pi^2} \left[\frac{2}{r^3} \sin(Gr/2) - \frac{G^2}{r^2} \cos(Gr/2) \right] \tag{4.14}$$

with j_1 being a spherical Bessel function shown in fig. 4.3.

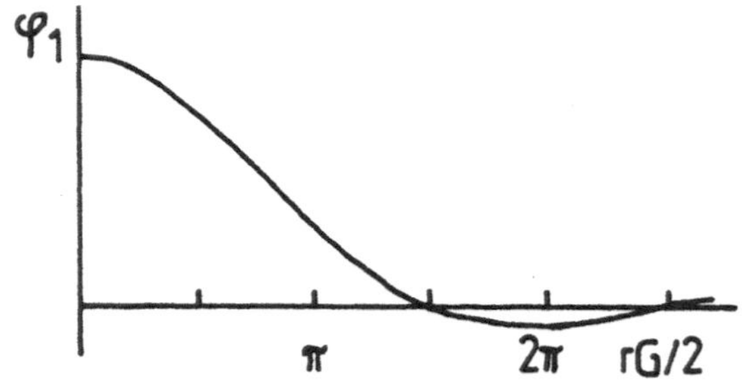

Fig. 4.3. Behaviour of the function φ_1 of (4.14).

The plain-wave expression (4.12) remains approximately valid for the Bloch wave-functions in the inner part of the Brillouin zone, if a weak crystal potential is present, whereas near the Brillouin zone boundaries terms $\exp[i(\mathbf{k}+\mathbf{G})\mathbf{r}]$ are now admixed where the reciprocal lattice vector $\mathbf{G}$ is such that $\mathbf{k}+\mathbf{G}$ is again near the Brillouin zone boundaries. This leads to an admixture of exponetials with $|\mathbf{k}|>G/2$ in the integral (4.13), with amplitudes decaying with increasing $|\mathbf{k}|$, and a corresponding change of the amplitudes for $|\mathbf{k}|<G/2$. Qualitatively this is achieved by extenting the integral (4.13) over the whole

extended **k**-space and introducing a step function $\Theta(G/2-|\mathbf{k}|)$ folded with a Gaussian of halfwidth q, that is, a factor

$$\int^{\infty} d^3z \, \frac{\Theta(G/2-z)}{(q\sqrt{2\pi})^3} \, e^{-(\mathbf{k}-\mathbf{z})^2/2q^2} \tag{4.15}$$

is to be insertad in the integral. The technical advantage of this procedure is that after the Fourier transformation it simply leads to a factor

$$\varphi_1(\mathbf{r}) \longrightarrow \varphi_1(\mathbf{r}) \, e^{-(rq)^2/2} \; . \tag{4.16}$$

The halfwidth q in (4.15) has the meaning of the depth of the region in **k**-space near the Brillouin zone boundary where plane-wave mixing takes place (fig. 4.4).

Treating the crystal potential by perturbation theory [122] yields

$$q \sim v_G/G \; , \quad v_G = \langle \mathbf{k}|v|\mathbf{k}+\mathbf{G}\rangle . \tag{4.17}$$

For nearly free electrons v_G is a typical pseudo-potential matrix element:

$$v_G \approx \frac{\varepsilon_F}{5} \approx \frac{1}{10}\left(\frac{G}{2}\right)^2 , \text{ d.h.} \quad q \approx G/40. \tag{4.18}$$

The corresponding Wannier function (4.16) is essentially localized in the range

$$r \leq \frac{2}{q} \approx \frac{80}{G} \approx 12R \tag{4.19}$$

with $R \approx 2\pi/G$ as the modulus of the smallest lattice vector. Applying these estimates in a crude qualitative manner to the tight-binding case $v_G \gg \varepsilon_F \approx \approx G^2/8$ again leads to the result that the Wannier function is in this case localized within the atom volume. A mathematically rigorous treatment of the problem, connecting the decay length of the Wannier function with the branch point $\mathbf{k}_0$ of the complex band structure, has led to a theorem by Kohn [73] confirming the issue of our simple intuitive considerations.

The result (4.19) might, however, only be regarged as an upper bound for the extension of the Wannier state, since it was obtained by arbitrarily fixing the phase factor $\alpha_\nu(\mathbf{k})\equiv 0$. An invariant lower bound for the spatial extension of the Wannier state, including also more complicated situations, is obtained from considering the asymptotics of (4.11). The analytic behaviour of the function $\varepsilon_\nu(\mathbf{k})$ determines the asymptotic behaviour of its Fourier components $\varepsilon_{\mathbf{R}\nu}$ for large $|\mathbf{R}|$. If $\varepsilon_\nu(\mathbf{k})$ is an analytic function of $\mathbf{k}$ which is always true for an isolated non-hybridized band [12,31], then $\varepsilon_{\mathbf{R}\nu}$ is majorized by any power of $1/|\mathbf{R}|$. Indeed, multiplying the second equation (4.8) with

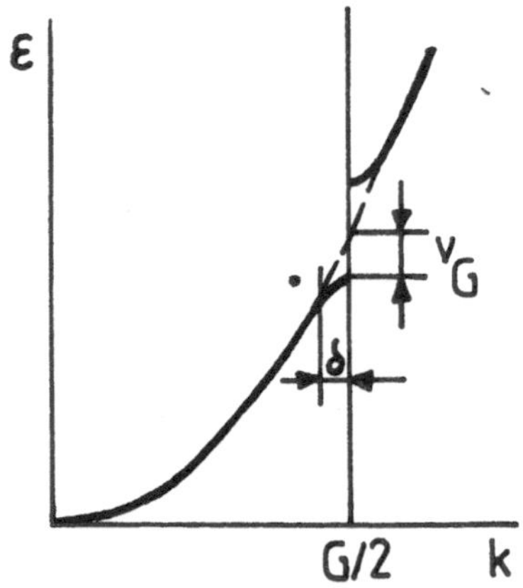

Fig. 4.4. Nearly-free electron band-structure near the zone boundary. In a range of depth $q \sim v_G/G$ the mixing of plane waves with wave-vectors $\mathbf{k}$ and $\mathbf{k+G}$ with $||\mathbf{k}|-|\mathbf{k+G}||<q$ takes place.

$$(\mathbf{R}^2)^m = e^{-i\mathbf{kR}} \left[\left(-i\frac{\partial}{\partial \mathbf{k}}\right)^2\right]^m e^{i\mathbf{kR}}$$

and integrating repeately by parts (taking into account the periodicity of $\varepsilon_\nu(\mathbf{k})$ which cancels the surface terms) we find that $(\mathbf{R}^2)^m\varepsilon_{\mathbf{R}\nu}$ is bounded together with the 2m-th derivate of $\varepsilon_\nu(\mathbf{k})$ as the derivative of an analytic periodic function.

If $\varepsilon_\nu(\mathbf{k})$ jumps on a closed surface in $\mathbf{k}$-space (as e.g. the flat hybridized band of fig. 4.4), then one finds (in analogy to (4.13) and (4.14)) the asymptotics $\varepsilon_{\mathbf{R}\nu}\sim|\mathbf{R}|^{-2}$, in the case of a bend of $\varepsilon_\nu(\mathbf{k})$ (as for the free electron band $\varepsilon_1(\mathbf{k})$ on the Brillouin zone boundary) $\varepsilon_{\mathbf{R}\nu}\sim|\mathbf{R}|^{-3}$ is found and so on. Be now

$$\varepsilon_{\mathbf{R}\nu} \sim |\mathbf{R}|^{-m} \qquad \text{for} \qquad |\mathbf{R}| \longrightarrow \infty \tag{4.20}$$

and

$$\{\mathbf{r}|0\nu\} \sim r^{-n} \qquad \text{for} \qquad r \longrightarrow \infty . \tag{4.21}$$

In this case, a rescaling $r \longrightarrow \alpha r$ yields for the matrix element

$$\{0\nu|\hat{h}|\mathbf{R}\nu\} \sim \{0\nu|\hat{t}|\mathbf{R}\nu\} \sim \alpha^3 \alpha^{-2} \alpha^{-2n} \ . \tag{4.22}$$

Together with $\varepsilon_{\mathbf{R}\nu} \sim \alpha^{-m}$, (4.11) implies now

$$n = (m+1)/2 \ . \tag{4.23}$$

Hence, for the isolated band (analytic $\varepsilon_\nu(\mathbf{k})$) we have again the exponential decay of the Wannier function. For the first free-electron band (m=3) we get back the $1/r^2$-decay of (4.14) and for the hybridized tight-binding band of fig. 4.2 (m=2) one would have n=3/2, that is the corresponding Wannier function would unavoidably be non-normalizable.

4.3. Multi-band Wannier functions

The just presented analysis shows in a simple manner that for a well separated isolated band the existence of a well localized Wannier function may be taken for granted, which function flows however out as the band gaps are decreased against the band width. But the analysis shows also that in the case of crossing bands with points of degeneracy (singular points of $\varepsilon_\nu(\mathbf{k})$) the one-band Wannier functions do no longer decay exponentially, are generally rather extended over many lattice sites, and are sometimes (for $N \to \infty$) even non-normalizable, that is the transformation (4.3) may prove not to be isometric. This difficulty appeared already in the classical work of Des Cloizeaux [25] as an incompleteness problem of symmetrically constructed Wannier functions. Lateron it was analized by Lix [91] and Krüger [77]. Lix actually asserts that this difficulty may be removed by an arbitrarily small change of the bands in a small surroundings of singular points, but it is clear from the above discussion that thereby only an arbitrarily large decay length of the Wannier function is gained.

The way out was already comprehensively discussed by Des Cloizeaux [25,26] and recently reanalized [77,74]. Another $\mathbf{k}$-dependent unitary matrix transformation $f_{\nu\mu}(\mathbf{k})$, $\mathbf{f}^{-1}=\mathbf{f}^+$ is inserted into (4.3) and (4.5) generalizing these transformations to

$$|\mathbf{R}\mu\} = \frac{1}{\sqrt{N}} \sum_{\mathbf{k}} \sum_{\nu} |\mathbf{k}\nu\rangle \, f_{\nu\mu}(\mathbf{k}) \, e^{-i\mathbf{k}\mathbf{R}} \ , \tag{4.24}$$

$$|\mathbf{k}\nu\rangle = \frac{1}{\sqrt{N}} \sum_{\mathbf{R}} \sum_{\mu} |\mathbf{R}\mu\} \, f^*_{\nu\mu}(\mathbf{k}) \, e^{i\mathbf{k}\mathbf{R}} \ . \tag{4.25}$$

The sums over ν and μ run over all bands in a hybridized band group. The equivalence of (4.6) and (4.7) (for finite N) holds as previously. (4.9) and (4.11) are to be replaced by

$$\hat{h}|0\nu\} = \sum_{\mathbf{R}} \sum_{\mu} |\mathbf{R}\mu\} \, \varepsilon_{\mathbf{R},\mu\nu} \quad \text{and} \quad \{\mathbf{R}\mu|\hat{h}|0\nu\} = \varepsilon_{\mathbf{R},\mu\nu} , \tag{4.26}$$

respectively, where the relations

$$\delta_{\mu\nu}\varepsilon_{\nu}(\mathbf{k}) = \sum_{\mathbf{R}} e^{-i\mathbf{k}\mathbf{R}} \sum_{\sigma\tau} f_{\mu\sigma}(\mathbf{k}) \, \varepsilon_{\mathbf{R},\sigma\tau} \, f^{*}_{\nu\tau}(\mathbf{k}) ,$$

$$\varepsilon_{\mathbf{R},\sigma\tau} = \frac{1}{N} \sum_{\mathbf{k}} e^{i\mathbf{k}\mathbf{R}} \sum_{\nu} f^{*}_{\nu\sigma}(\mathbf{k}) \, \varepsilon_{\nu}(\mathbf{k}) \, f_{\nu\tau}(\mathbf{k}) . \tag{4.27}$$

replace (4.8).

It can be formally shown [26] that the states (4.24) again decay exponentially, if the considered band group is isolated from the remaining band structure. This has to do with the fact that the non-ordered set $\{\varepsilon_{\nu}(\mathbf{k}) | \nu \in \text{band group}\}$ always is an analytic set-valued function of $\mathbf{k}$ in the sense that for any analytic curve $\mathbf{k}(t)$ a numbering exists (dependend on the curve considered) for which $\varepsilon_{\nu}(\mathbf{k}(t))$ are analytic functions of t [31]. This guarantees the existence of a unitary transformation depending (in a non-analytic way in points of band crossing) on $\mathbf{k}$ and such that the functions

$$\varepsilon_{\sigma\tau}(\mathbf{k}) = \sum_{\nu} f^{*}_{\nu\sigma}(\mathbf{k}) \, \varepsilon_{\nu}(\mathbf{k}) \, f_{\nu\tau}(\mathbf{k}) \tag{4.28}$$

are individually analytic functions of $\mathbf{k}$ and hence their Fourier components $\varepsilon_{\mathbf{R},\sigma\tau}$ decay exponentially. These latter Fourier components must be identified with the entities in squared brackets of (4.1) thus justifying the ansatz with a finite lattice sum containing a small number of items in the LCAO interpolation schemes. But a quantitative statement about the decay length can hardly be made.

From all that it follows that, apart from the elementary case of an isolated tightly bound s-band, Wannier functions are extremely complicated beings. An attempt to calculate them directly from the Bloch functions via (4.24) [19] seems not to be very encouraging. Another interesting approach is that of Chadi and Cohen [22]. However, again most helpful seems to be a variational principle. It was long ago stated independendly by Koster [76] and Parzen [101] and was recently used by Kohn [74] for a practical computational scheme. The band structure energy (tr means the trace of a matrix)

$$E = \sum_{\mathbf{k}\nu} \varepsilon_\nu(\mathbf{k}) = \sum_{\mathbf{k}} \operatorname{tr}(\delta_{\mu\nu}\varepsilon_\nu(\mathbf{k})) \tag{4.29}$$

of a band group may be considered as the ground-state energy of a system of non-interacting particles moving in the crystal field and is therefore stationary against variations of the states $|\mathbf{k}\nu\rangle$ under the constraint (4.6). From (4.26), (4.27) and with the unitary of $\mathbf{f}$ we may write

$$E = \sum_{\mathbf{R}} \sum_{\mathbf{k}} e^{-i\mathbf{k}\mathbf{R}} \operatorname{tr} \varepsilon_{\mathbf{R},\sigma\tau} = \operatorname{tr} \varepsilon_{0,\sigma\tau} = \sum_{\mu} \{0\mu|\hat{h}|0\mu\} \tag{4.30}$$

and have the variation principle

$$\delta \sum_{\mu} \{0\mu|\hat{h}|0\mu\} = 0 \;, \quad \{\mathbf{R}'\mu|\mathbf{R}\nu\} = \delta_{\mu\nu}\,\delta_{\mathbf{R}'\mathbf{R}} \;. \tag{4.31}$$

Instead of (4.6) the equivalent constraint (4.7) was added in (4.31). The variation immediately gives back (4.26) with the $\varepsilon_{\mathbf{R},\sigma\tau}$ as Lagrange multipliers. The difficulty is that in the general case nothing is known about how to choose the transformation $\mathbf{f}$ in order to obtain well localized Wannier functions. This brings a great uncertainly into the choice of the trial functions for the variation. In a few simple cases the problems may be solved [74,77]. The solution of (4.31) is further complicated by the form of the constraints. Therefore Kohn [74] proposed to choose appropriate local trial functions $|\mathbf{R}n)$, to make up Bloch states (2.50) of them, to orthonormalize the latter for each $\mathbf{k}$ separately via Cholesky's procedure, and to transform the orthonormalized trial Bloch states into trial Wannier states via (4.3) which latter now already fulfil the constraint of (4.31). This programme was acomplished by Andreoni [5].

The same procedure may be used to build up orthonormalized local basis states out of atom-like basis states: Take the non-orthogonal Bloch states (2.50) (for a given $\mathbf{k}$) and orthonormalize them with the help of the overlap matrix (2.53),

$$|\mathbf{k}\mathbf{S}n\} \equiv \sum_{\mathbf{S}'m} |\mathbf{k}\mathbf{S}'m\rangle\, D^{-1}_{\mathbf{S}'m;\mathbf{S}n}(\mathbf{k}) \;, \tag{4.32}$$

where the matrix $\mathbf{D}$ is again the square root of the matrix $\mathbf{B}=\mathbf{D}^T\mathbf{D}$ of (2.53) according to Cholesky's procedure. (In mathematics (4.32) is called a Schmidt procedure.) Due to (2.45) the states (4.32) are also orthogonal to all core states. The orthonormal local states are then obtained by a Fourier transformation of (4.32):

$$|\mathbf{R}+\mathbf{S}n\} = \frac{1}{\sqrt{N}} \sum_{\mathbf{k}} |\mathbf{k}\mathbf{S}n\} \, e^{-i\mathbf{k}(\mathbf{R}+\mathbf{S})} \; . \tag{4.33}$$

The LCAO representation (2.51) of the band states is now replaced by

$$|\mathbf{k}\nu\rangle = \sum_{\mathbf{S}n} |\mathbf{k}\mathbf{S}n\} \, f^{*}_{\nu,\mathbf{S}n}(\mathbf{k}) \tag{4.34}$$

with the unitary matrix $\mathbf{f}^{+}=\mathbf{DC}$ and $\mathbf{C}$ being the coefficients matrix of (...51). Inserting here the relation inverse to (4.33) the representation (4.34) becomes formally idendical with (4.25) (up to a constant phase factor $\exp(i\mathbf{kS})$ which might be included in the unitary matrix $\mathbf{f}^{+}$). Inserting into the inverse relation (4.24) the LCAO representation (2.50-51) of $|\mathbf{k}\nu\rangle$ the connection between the orthonormal local states (4.33) and the atom-like states $|\mathbf{R}+\mathbf{S}n\rangle$ entering (2.50) is found to be

$$|\mathbf{S}'m\} = \sum_{\mathbf{R}\mathbf{S}n} |\mathbf{R}+\mathbf{S}n\rangle \, a^{(\mathbf{R})}_{\mathbf{S}n;\mathbf{S}'m} \; , \tag{4.35}$$

$$a^{(\mathbf{R})}_{\mathbf{S}n;\mathbf{S}'m} = \frac{1}{N} \sum_{\mathbf{k}} e^{i\mathbf{k}(\mathbf{R}+\mathbf{S}-\mathbf{S}')} \, [D^{-1}(\mathbf{k})]_{\mathbf{S}n;\mathbf{S}'m} \; . \tag{4.36}$$

Up to now there has been left out of the consideration one important point: Maintaining the analogy of the obtained relations with those relations (4.24-25) for Wannier functions it is implicitly assumed that (2.51) be an <u>exact</u> representation of the true band states $|\mathbf{k}\nu\rangle$. More precisely it is assumed that (2.51) for a given $\mathbf{k}$ be a transformation from a number, say m, of states $|\mathbf{k}\mathbf{S}n\rangle$ to m true band states $|\mathbf{k}\nu\rangle$. This is generally not the case, and therefore the states (4.35) are only trial Wannier states. Parametrizing now the atom-like states $|\mathbf{R}+\mathbf{S}n\rangle$ and varying the first expression (4.31) would just be Kohn's procedure. However, as we have seen, this is equivalent to finding the minimum of the bandstructure energy E of (4.29). Recall that our optimum atom-like basis were actually just found by minimizing this bandstructure energy. Hence, inserting these optimum atom-like states into (4.35) is just equivalent to Kohn's procedure where the variation now has only been carried out a step earlier.

An alternative approach to treat the orthonormality constraints in the variational principle (4.31) was applied to the diamond lattice by Kane and Kane [68].

4.4 An example

For illustration, results for multi-band Wannier functions of the copper bandstructure [33] ar presented here. From the atom-like basis functions of the LCAO calculation of [55] the coefficients (4.36) were computed via (2.53). The nine atom-like basis functions were ordered with the localization decreasing, that is the index n of (4.35) and (4.36) running from 1 to 9 labels the states with angular dependences $3z^2-r^2$, x^2-y^2, yz, xz, xy, s, x, y, z in turn. In this way, the upper triangular form of the matrix $\mathbf{D}^{-1}$, carried over to the matrices $\mathbf{a}^{(\mathbf{R})}$, guarantees that for a pair of functions at a distance **R** the orthogonalization correction is always attached to the less localized function. Here are the coefficient matrices $\mathbf{a}^{(\mathbf{R})}$:

$$\mathbf{a}^{(000)} = \begin{pmatrix} 1. & & & & & & & & \\ & 1. & & & & & & & \\ & & 1. & & & & & & \\ & & & 1. & & & & & \\ & & & & 1. & & & & \\ & & & & & 1.055 & & & \\ & & & & & & 1.137 & & \\ & & & & & & & 1.219 & \\ & & & & & & & & 1.251 \end{pmatrix}$$

$$\mathbf{a}^{(011)} = \begin{pmatrix} 0.002 & 0.006 & -0.004 & 0. & 0. & 0.011 & -0.004 & 0.006 & 0. \\ & -0.001 & 0.006 & 0. & 0. & -0.019 & 0.031 & -0.012 & 0. \\ & & -0.007 & 0. & 0. & 0.035 & -0.046 & -0.045 & 0. \\ & & & 0.002 & 0.004 & 0. & 0. & 0. & 0.002 \\ & & & & 0.002 & 0. & 0. & 0. & 0.002 \\ & & & & & -0.047 & 0.131 & 0.138 & 0. \\ & & & & & & 0.077 & 0.208 & 0. \\ & & & & & & & 0.084 & 0. \\ & & & & & & & & 0.024 \end{pmatrix}$$

$$\mathbf{a}^{(002)} = \begin{pmatrix} 0. & 0. & 0. & 0. & 0. & 0. & 0. & 0. & 0. \\ & 0. & 0. & 0. & 0 & 0. & 0. & 0. & 0. \\ & & 0. & 0 & 0. & 0. & 0. & 0.004 & 0. \\ & & & 0. & 0. & 0. & 0. & 0. & 0.001 \\ & & & & 0. & 0. & 0. & 0. & 0. \\ & & & & & 0.009 & 0.004 & 0. & 0. \\ & & & & & & 0.013 & 0. & 0. \\ & & & & & & & -0.003 & 0. \\ & & & & & & & & 0.002 \end{pmatrix}$$

$$
\mathbf{a}^{(112)} = \begin{pmatrix}
0. & 0. & 0. & 0. & 0. & -0.001 & 0.002 & 0.001 & 0. \\
 & 0. & 0. & 0. & 0. & 0. & 0. & -0.004 & 0.003 \\
 & & 0. & 0. & 0. & -0.002 & 0.001 & 0. & -0.003 \\
 & & & 0. & 0. & -0.002 & 0.001 & -0.004 & 0. \\
 & & & & 0. & 0. & 0. & 0. & 0. \\
 & & & & & 0.006 & -0.005 & 0.011 & 0.009 \\
 & & & & & & -0.001 & 0.012 & 0.010 \\
 & & & & & & & 0.004 & 0.014 \\
 & & & & & & & & 0.001
\end{pmatrix}
$$

$$
\mathbf{a}^{(022)} = \begin{pmatrix}
0. & 0. & 0. & 0. & 0. & -0.001 & 0.002 & 0.002 & 0. \\
 & 0. & 0. & 0. & 0. & 0.001 & -0.001 & 0.002 & 0. \\
 & & 0. & 0. & 0. & -0.002 & 0.002 & -0.006 & 0. \\
 & & & 0. & 0. & 0. & 0. & 0. & 0.001 \\
 & & & & 0. & 0. & 0. & 0. & 0.001 \\
 & & & & & 0.003 & -0.004 & 0.020 & 0. \\
 & & & & & & -0.002 & 0.024 & 0. \\
 & & & & & & & 0.018 & 0. \\
 & & & & & & & & -0.002
\end{pmatrix}
$$

$$
\mathbf{a}^{(013)} = \begin{pmatrix}
0. & 0. & 0. & 0. & 0. & 0. & 0. & 0. & 0. \\
 & 0. & 0. & 0. & 0. & 0. & 0. & 0. & 0. \\
 & & 0. & 0. & 0. & 0. & 0. & 0. & 0. \\
 & & & 0. & 0. & 0. & 0. & 0. & 0. \\
 & & & & 0. & 0. & 0. & 0. & 0. \\
 & & & & & -0.001 & 0.002 & 0.001 & 0. \\
 & & & & & & 0.001 & 0.001 & 0. \\
 & & & & & & & 0.001 & 0. \\
 & & & & & & & & 0.
\end{pmatrix}
$$

$$
\mathbf{a}^{(222)} = \begin{pmatrix}
0. & 0. & 0. & 0. & 0. & 0. & 0. & 0. & 0. \\
 & 0. & 0. & 0. & 0. & 0. & 0. & 0. & 0. \\
 & & 0. & 0. & 0. & 0. & 0. & 0. & 0. \\
 & & & 0. & 0. & 0. & 0. & 0. & 0. \\
 & & & & 0. & 0. & 0. & 0. & 0. \\
 & & & & & -0.001 & 0.002 & 0. & 0.002 \\
 & & & & & & 0.001 & 0.001 & 0.002 \\
 & & & & & & & 0. & 0.003 \\
 & & & & & & & & 0.001
\end{pmatrix}
$$

For higher than the sixth neighbours all coefficients are smaller than 0.001. In fig. 4.5 and fig. 4.6 the corresponding 3d-derivative and 4s-derivative Wannier functions are plotted along some directions. As is seen, the 3d-derivative Wannier function is essentially equal to the atom-like basis function and is well localized, whereas the 4s-derivative Wannier function has rather large contributions at the first neighbour and very small contributions at the second neighbour.

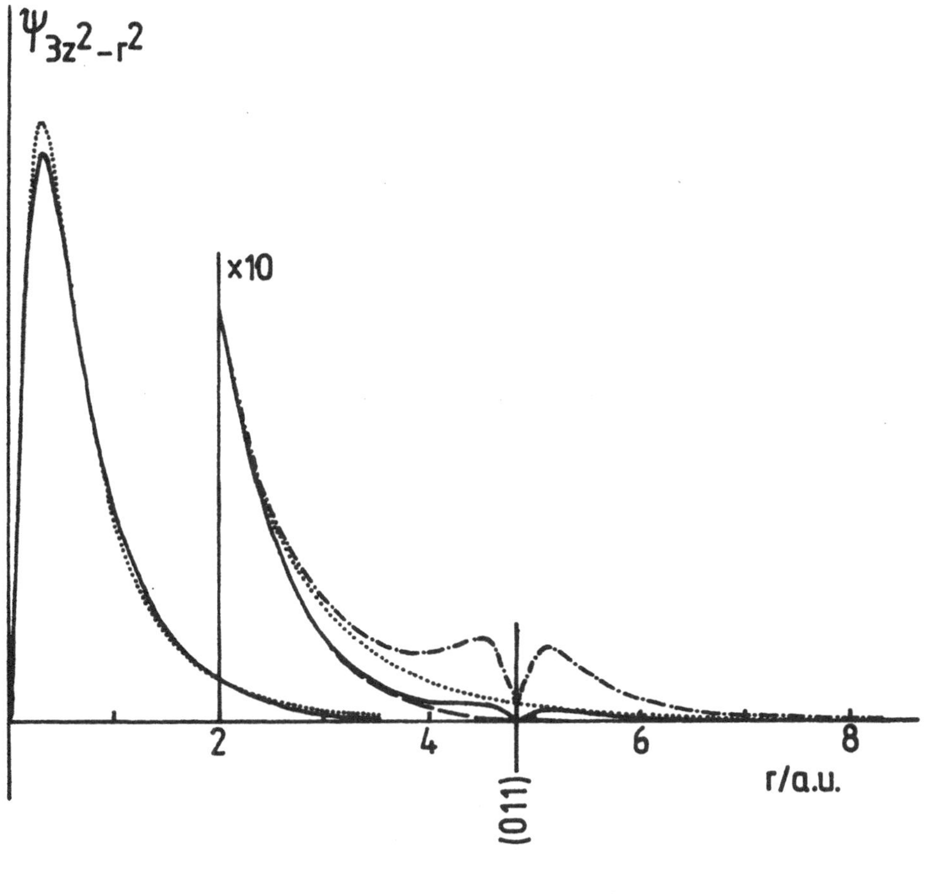

Fig. 4.5. 3d-derivative Wannier function of Cu (corresponding to the $a_{11}^{(R)}$ elements) along the (011)-direction (full curve). Also the corresponding basis function (broken curve) is shown and, for comparison, the free-atom 3d-function (dotted curve) and the Wannier function constructed from it (dot-dashed curve).

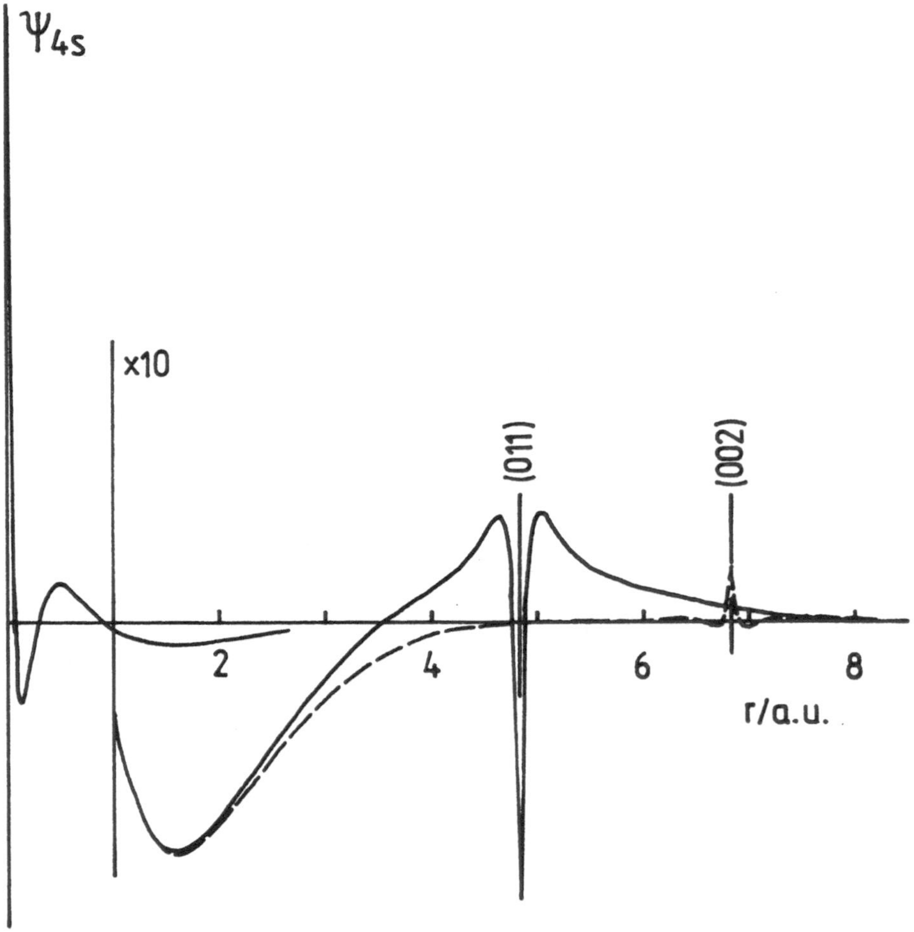

Fig. 4.6. 4s-derivative Wannier function of Cu along the (011)-direction (full curve) and along the (001)-direction (broken curve).

4.5. The recursion method for resolvent operator matrix elements

The LCAO interpolation of a bandstructure (for the sake of simplicity we consider the case of a single band as in the numerical example of section 4.1.)

$$\varepsilon_\nu(\mathbf{k}) = \sum_{\mathbf{R}} \varepsilon_{\mathbf{R}\nu} \; e^{-i\mathbf{kR}} = \lim_{R\to\infty} \frac{\sum_{\mathbf{R}}^{|\mathbf{R}|<R} h_{\mathbf{R}\nu} \; e^{-i\mathbf{kR}}}{\sum_{\mathbf{R}}^{|\mathbf{R}|<R} b_{\mathbf{R}\nu} \; e^{-i\mathbf{kR}}}$$

is an example of the very common situation, where the last expression may converge much faster than the series of the second expression (cf. (4.2)). Given a power series

$$f(x) = \sum_{k=0}^{\infty} c_k x^k , \tag{4.37}$$

the closest approximation of a rational fraction

$$f_{[N,M]}(x) = \left[\sum_{k=0}^{N} a_k x^k\right] \Big/ \left[\sum_{k=0}^{M} b_k x^k\right] \tag{4.38}$$

to $f(x)$ is called its [N,M] Padé approximant [18]. On the background of the discussion of section 4.1. it would be a challenging task to find an LCAO interpolation scheme in terms of Padé approximants.

Another problem leading directly to Padé approximants is that of calculating resolvent operator matrix elements. A wide class of physical observables may be expressed in terms of such matrix elements [70,115], a simple example being the orbital projected density of states

$$D_\varphi(\varepsilon) = -\frac{1}{\pi} \operatorname{Im}\langle\varphi|[\varepsilon^+ - \hat{h}]^{-1}|\varphi\rangle , \quad \varepsilon^+ \equiv \varepsilon + i0 . \tag{4.39}$$

Denoting the resolvent operator by

$$\hat{g}(\varepsilon) = (\varepsilon - \hat{h})^{-1}, \tag{4.40}$$

its general matrix element may be expressed by the Liouville-Neumann series

$$w(\varepsilon) = \langle\psi|\hat{g}(\varepsilon)|\varphi\rangle = \frac{1}{\varepsilon}\left[\langle\psi|\varphi\rangle + \frac{1}{\varepsilon}\langle\psi|\hat{h}|\varphi\rangle + \frac{1}{\varepsilon^2}\langle\psi|\hat{h}^2|\varphi\rangle + \dots\right] \equiv$$

$$\equiv \frac{1}{\varepsilon}\left[w_0 + \frac{1}{\varepsilon}w_1 + \frac{1}{\varepsilon^2}w_2 + \dots\right] \tag{4.41}$$

Alternatively the last sum may via

$$w(\varepsilon) = \langle\psi|\chi\rangle \tag{4.42}$$

be obtained from the solution χ of the inhomogeneous equation

$$(\varepsilon - \hat{h})\chi = \varphi , \tag{4.43}$$

the Liouville-Neumann series of the latter solution being

$$\chi = \frac{1}{\varepsilon}\left[\varphi + \frac{1}{\varepsilon}\hat{h}\varphi + \frac{1}{\varepsilon^2}\hat{h}^2\varphi + \ldots\right] \equiv$$

$$\equiv \frac{1}{\varepsilon}\left[\varphi_0 + \frac{1}{\varepsilon}\varphi_1 + \frac{1}{\varepsilon^2}\varphi_2 + \ldots\right] . \tag{4.44}$$

After introducing another sequence of states $\psi_k = (\hat{h}^+)^k\psi$ and the matrix $\mathbf{B} = [[\langle\psi_i|\varphi_j\rangle]]$,

$$\hat{P}_N = \sum_{i,j=1}^{N} |\varphi_i\rangle[\mathbf{B}^{-1}]_{ij}\langle\psi_j| \tag{4.45}$$

defines the (in general non-orthogonal) projection onto the (N+1)-dimensional state space S_N spanned by the φ_i, $i=0,\ldots,N$ from the (N+1)-dimensional state space S_N' spanned by the ψ_i. (Fig. 4.7. For $\varphi = \psi$ and $\hat{h} = \hat{h}^+$ the projection is orthogonal.) The (N+1)-dimensional truncation of the problem (4.43) is now (recall $\varphi = \varphi_0 \in S_N$)

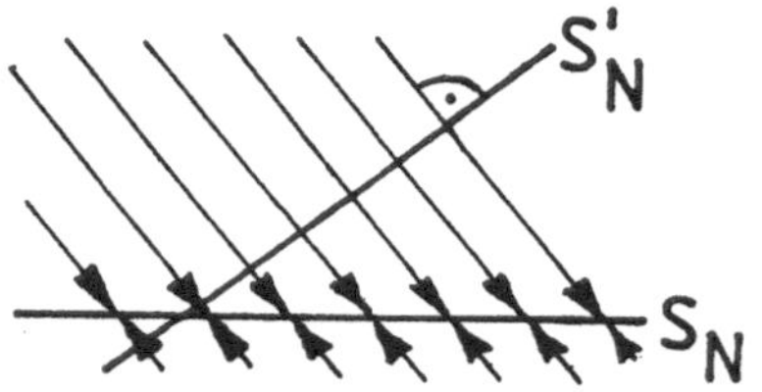

Fig. 4.7. The projection $\hat{P}_N$

$$(\varepsilon - \hat{P}_N\hat{h}\hat{P}_N)\chi^{(N)} = \varphi , \tag{4.46}$$

and its solution may exactly be represented by the finite series

$$\chi^{(N)} = \sum_{j=0}^{N} a_j\varphi_j \tag{4.47}$$

the coefficients of which are easily obtained by inserting (4.47) into (4.46). The corresponding approximant to (4.42) is

$$w^{(N)}(\varepsilon) = \langle\psi|\chi^{(N)}\rangle = \sum_{j=0}^{N} a_j(\varepsilon)\, w_j \tag{4.48}$$

with the w_j from (4.41). As the $a_j(\varepsilon)$ are obtained from a (N+1)-dimensional linear equation system the coefficients of which are linear functions of ε, they appear to be rational polynomial expressions in ε of the order of at most N in the numerator and of at most (N+1) in the denominator (which is most easily seen from Kramer's determinantal solution of the linear equation system). It can be shown [99] that (4.48) is just the [N,N+1] Padé approximant of (4.41).

If the $\varphi_k \equiv (\hat{h})^k\varphi$ were orthogonal to each other the expansion (4.44) would be uniquely determined, and the comparison with (4.47) would yield $a_j(\varepsilon) = \varepsilon^{-j-1}$. However, for a Hermitean $\hat{h}$ with $\hat{h} \neq 0$ the φ_k may never be orthogonal to each other which is seen at once from the contradiction

$$0 = \langle\varphi_o|\varphi_2\rangle = \langle\varphi_o|\hat{h}|\varphi_1\rangle = \langle\varphi_1|\varphi_1\rangle \neq 0.$$

Therefore, the expansion (4.44) is never unique, and in a wide class of cases the expansion (4.47) converges much faster then (4.44); it may even converge in cases where (4.44) does not; likewise for (4.41) and (4.48).

The Padé approximants (4.48) for the special case $\psi = \varphi$ are most conveniently expressed in terms of continued fractions. The first step on this way is again the introduction of an orthogonalized basis $\{|i\rangle\}$ instead of the φ_i, so that the first (N+1) basis states $|0\rangle,\ldots,|N\rangle$ span the same state space S_N as the $|\varphi_o\rangle,\ldots,|\varphi_N\rangle$ do (Schmidt procedure). Then, because $\hat{h}S_i \subset S_{i+1}$, the matrix elements $\langle j|\hat{h}|i\rangle$ are zero for $j>i+1$ (upper Hessenberg form). Since $\hat{h}$ is Hermitean, the matrix $\langle j|\hat{h}|i\rangle$ is in fact tridiagonal. Hence

$$b_{i+1}|i+1\rangle = \hat{h}|i\rangle - a_i|i\rangle - b_i^*|i-1\rangle , \tag{4.49}$$

where

$$|0\rangle = |\varphi\rangle, \quad b_o = 0, \quad a_i = \langle i|\hat{h}|i\rangle, \tag{4.50}$$

and

$$b_{i+1} = \langle i+1|i+1\rangle^{-1/2} \tag{4.51}$$

is obtained from (4.49). These equations determine the matrix elements

$$[[\langle i|\varepsilon - \hat{h}|j\rangle]] = \begin{pmatrix} \varepsilon-a_o & -b_1^* & 0 & 0 & \cdots \\ -b_1 & \varepsilon-a_1 & -b_2^* & 0 & \cdots \\ 0 & -b_2 & \varepsilon-a_2 & -b_3^* & \cdots \\ 0 & 0 & -b_3 & \varepsilon-a_3 & \cdots \\ \vdots & \vdots & \vdots & \vdots & \ddots \end{pmatrix}$$

recursively. Defining the determinants

$$D = \det[[\langle i|\varepsilon - \hat{h}|j\rangle]], \quad D_l = \det[[\langle i|\varepsilon - \hat{h}|j\rangle_{i,j>l}]] \tag{4.52}$$

the matrix element of the resolvent is

$$w(\varepsilon) = \langle\varphi|\hat{g}(\varepsilon)|\varphi\rangle = \langle 0|[\varepsilon - \hat{h}]^{-1}|0\rangle =$$

$$= D_o/D = \frac{1}{\varepsilon - a_o - |b_1|^2 D_1/D_0} =$$

$$= \cfrac{1}{\varepsilon - a_o - \cfrac{|b_1|^2}{\varepsilon - a_1 - \cfrac{|b_2|^2}{\varepsilon - a_2 - \dots}}} . \tag{4.53}$$

The restriction to the state space S_N now means truncating the matrix $[[\langle i|\varepsilon - \hat{h}|j\rangle]]$, and hence the Padé approximant (4.48) is obtained truncating the continued fraction (4.53) by putting

$$b_{N+1} = 0 . \tag{4.54}$$

This machinery was first applied to the resolvent of tight-binding Hamiltonians by Haydock et al. [53].

Up to now, our recursion equation (4.49) is still a functional one. It may be algebraized by using a matrix representation of the form (4.26)

$$\hat{h} = \sum_{\mathbf{R}'\mu} \sum_{\mathbf{R}\nu} |\mathbf{R}'\mu\} \varepsilon_{\mathbf{R}'-\mathbf{R},\mu\nu} \{\mathbf{R}\nu| \tag{4.55}$$

for the Hamiltonian. Of course the full benefit of the method can only be reaped if the matrix $\varepsilon_{\mathbf{R}'-\mathbf{R},\mu\nu}$ is sufficiently small with respect to $\mathbf{R}'-\mathbf{R}$ because otherwise the states $|i\rangle$ blow up in real space very rapidely. Instead of the Wannier representation (4.55) a sufficiently complete ordinary LCAO representation

$$\hat{h} = \sum_{\mathbf{R}'\mathbf{S}'m} \sum_{\mathbf{RS}n} |\mathbf{R}'\mathbf{S}'m\rangle[\mathbf{B}^{-1}\mathbf{H}\mathbf{B}^{-1}]_{\mathbf{R}'-\mathbf{R};\mathbf{S}'m,\mathbf{S}n}\langle\mathbf{RS}n| \tag{4.56}$$

could be used likewise, but here the matrix $\mathbf{B}^{-1}\mathbf{H}\mathbf{B}^{-1}$ is only small for an optimally localized LCAO basis which was shown in section 4.3. to be largely equivalent to a Wannier basis. Otherwise the factors $\mathbf{B}^{-1}$ may drastically enlarge the spatial extension of the matrix elements of $\hat{h}$ (with respect to $\mathbf{R}'-\mathbf{R}$). ((4.56) is build in analogy to (4.46) with the orthogonal projector onto the state space spanned by atom-like orbitals.)

Introducing the continued fraction representation (4.53) is not only a convenient way of explicitly writing down the Padé approximants. By its means a great variety of further partial summations of the series (4.41) can be produced. For instance, instead of (4.54),

$$a_i = a_N \; , \quad b_{i+1} = b_{N+1} \qquad \text{for } i>N \tag{4.57}$$

may be used. For this assumption the end of the continued fraction may easily be summed up:

$$\cfrac{1}{\varepsilon - a_N - \cfrac{|b_{N+1}|^2}{\varepsilon - a_N - \cfrac{|b_{N+1}|^2}{\varepsilon - a_N - \dots}}} = t_N(\varepsilon) = \frac{1}{\varepsilon - a_N - |b_{N+1}|^2 t_N(\varepsilon)} \; , \tag{4.58}$$

and hence

$$t_N(\varepsilon) = \frac{1}{2|b_{N+1}|^2}\left[\varepsilon - a_N - \sqrt{(\varepsilon - a_N)^2 - 4|b_{N+1}|^2}\,\right] . \tag{4.59}$$

This may be inserted into the N-th denominator of (4.53). There are many other possibilities [86]. Of course, as for any partial summation, physical intuition is needed in order to get a reasonable answer.

Any sequence (a_i, b_i); $i=N+1, N+2, \dots$ like (4.57), obtained independently of the recursion equations (4.49–51), is called a terminator of the continued fraction (4.53). In accordance with (4.39), the orbital projected density of states of the terminator is

$$D_N(\varepsilon) = -\frac{1}{\pi}\,\mathrm{Im}\, t_N(\varepsilon) \; , \tag{4.60}$$

where $t_N(\varepsilon)$ is the value of the continued fraction of the terminator as in (4.58). In this context the possibility to invert (4.53) and to construct the sequence (a_i, b_i); $i=0,1,\dots$, that is, the tridiagonal Hamiltonian from a given orbital projected density of states $-(1/\pi)\mathrm{Im}\,w(\varepsilon)$ is very challenging [55]. For instance, the sequence (a_i, b_i) might be determined from the given density of states of a pure crystal, and its terminator $t_N(\varepsilon)$ could then be used in the continued fraction corresponding to a locally perturbed structure. It should be recognized that the procedure described in the paper [55] establishes another local orbital interpolation scheme for the electronic structure.

4.6. The chemical pseudo-potential

P. W. Anderson's chemical-pseudo-potential method [3,4] may be characterized of holding an intermediate position between non-orthogonal basis representations and Wannier representations. Though the pseudo-wave functions are non-orthogonal, yet an ordinary eigenvalue problem $(\mathbf{D}-\varepsilon\mathbf{I})\mathbf{C} = 0$ with the unit matrix $\mathbf{I}$ instead of a generalized on $(\mathbf{H}-\varepsilon\mathbf{S})\mathbf{C} = 0$ as in (2.49) or (2.52) is obtained. However, the matrix $\mathbf{D}$ is non-Hermitean. But for a properly chosen pseudo-potential the pseudo-wavefunctions may be even more localized than Wannier functions [3]. The starting-point is the pseudo-potential equation for the atom site $\mathbf{R}$

$$(-\frac{\Delta}{2} + v_{\mathbf{R}})|\mathbf{R}m\rangle + \left[v'_{\mathbf{R}} - \sum_{\mathbf{R}'(\neq\mathbf{R})} \sum_{n} |\mathbf{R}'n\rangle\langle\mathbf{R}'n|0_{\mathbf{R}}\right]|\mathbf{R}m\rangle = \eta_m|\mathbf{R}m\rangle \tag{4.61}$$

where the potential representation $v = \sum v_{\mathbf{R}}$ is used, and

$$v'_{\mathbf{R}} = v - v_{\mathbf{R}} \, . \tag{4.62}$$

The square brackets of (4.61) contain the chemical pseudo-potential with $0_{\mathbf{R}}$ being any suitable operator though in practice only $0_{\mathbf{R}} = v'_{\mathbf{R}}$ has been tried. The m-th pseudo-state localized at $\mathbf{R}$ is $|\mathbf{R}m\rangle$. The equation (4.61) must be solved self-consistently as it is non-linear in the pseudo-states $|\mathbf{R}m\rangle$. A perturbation treatment starting with atom states is given in [4]. With the use of (4.61) and (4.62), the result of operating with $\hat{h} = -\Delta/2 + v$ onto the pseudo-state $|\mathbf{R}m\rangle$ is

$$\hat{h}|\mathbf{R}m\rangle = \eta_m|\mathbf{R}m\rangle + \sum_{\mathbf{R}'(\neq\mathbf{R})} \sum_{n} |\mathbf{R}'n\rangle\langle\mathbf{R}'n|0_{\mathbf{R}}|\mathbf{R}m\rangle = \sum_{\mathbf{R}'n} |\mathbf{R}'n\rangle \, K_{\mathbf{R}'n;\mathbf{R}m} \tag{4.63}$$

defining the non-Hermitean matrix $\mathbf{K}$. The ansatz

$$|\nu) = \sum_{\mathbf{R}m} |\mathbf{R}m\rangle \, c_{\mathbf{R}m}; \tag{4.64}$$

for the solution of the Schrödinger equation

$$\hat{h}|\nu) = |\nu)\varepsilon_{\nu} \tag{4.65}$$

now leads to

$$(\hat{h} - \varepsilon_\nu)|\nu\rangle = \sum_{\mathbf{R}m} \sum_{\mathbf{R}'n} |\mathbf{R}'n\rangle \, (K_{\mathbf{R}'n;\mathbf{R}m} - \varepsilon_\nu \delta_{\mathbf{R}'\mathbf{R}} \delta_{nm}) \, c_{\mathbf{R}m;\nu} = 0 \ . \quad (4.66)$$

Because of the linear independence of the state vectors $|\mathbf{R}'n\rangle$ the last equation results in

$$\boxed{\sum_{\mathbf{R}m} (K_{\mathbf{R}'n;\mathbf{R}m} - \varepsilon_\nu \delta_{\mathbf{R}'\mathbf{R}} \delta_{nm}) \, c_{\mathbf{R}m;\nu} = 0} \quad (4.67)$$

for all $\mathbf{R}'$ and n. This is the mentioned ordinary eigenvalue problem for a non-Hermitean matrix **K**. It was originally introduced by P. W. Anderson as a theoretical basis of the Hückel theory of π-electron systems in conjugated polymer chains. Since the chemical pseudo-potential of (4.61) can be made rather weak by a suitable choice of $\hat{O}_\mathbf{R}$, in many applications atom states may be used as approximate pseudo-states. However, there is no benefit in neglecting the chemical pseudo-potential completely. Putting the expression in the square brackets of (4.61) equal to zero would mean .

$$\sum_{\mathbf{R}'(\neq\mathbf{R})} \sum_{n} |\mathbf{R}'n\rangle\langle\mathbf{R}'n|\hat{O}_\mathbf{R}|\mathbf{R}m\rangle = v'_\mathbf{R}|\mathbf{R}m\rangle =$$

$$= \sum_{\mathbf{R}'n} \sum_{\mathbf{R}''p} |\mathbf{R}'n\rangle[\mathbf{B}^{-1}]_{\mathbf{R}'n;\mathbf{R}''p}\langle\mathbf{R}''p|v'_\mathbf{R}|\mathbf{R}m\rangle$$

which would essentially result in $\mathbf{K} = \mathbf{B}^{-1}\mathbf{H}$ with the same problems as in (4.56). (In the last equality, again an orthogonal projector of the type (4.46) has been introduced as in (4.56) in order to show the algebraic structure of the functional equation.) Only in the tight-binding case one would , of course, get back $\mathbf{K} = \mathbf{H}$. It is just the presence of the non-zero chemical pseudo-potential which reduces the spatial extension of the pseudo-states and thus the extension of the matrix **K** with respect to $\mathbf{R}'-\mathbf{R}$. In the general case the similarity of the equations (4.67) with the tight-binding equations favours the use of the recursion method described in the last section. However, because of the non-Hermiticity of **K** its Hessenberg form is not necessarily tridiagonal (though Hermiticity is not a necessary condition for the Hessenberg form to be tridiagonal), and the consequences of neglecting this fact are not very clear. A survey over applications may be found in [15].

Very recently an alternative optimum basis LCAO approach has been based on a generalization of the philosophy underlying the content of this section [61].

5. THE LOCAL BASIS REPRESENTATIONS OF THE ELECTRON DENSITY

This and the next chapters are concerned with closing the loop of one iteration step for a self-consistent solution of (1.26) after having performed the band structure calculation as described in chapters 2. and 3. with the potential $v(\mathbf{r})$ of that iteration step. The first part of the iteration step, the bandstructure calculation for a given potential $v(\mathbf{r})$, was distinguished by its great conceptual and numerical simplicity and by its comprehensibility in chemical terms. To be honest it must be admitted that, because of the unavoidably appearing double lattice sums, much of the numerical simplicity is lost if the electron density is to be calculated from LCAO wavefunctions. Indeed, the calculation of the multi-centre integrals (section 2.4.) and the calculation of the total electron density described in this chapter are the two time-consuming steps of a self-consistent LCAO bandstructure calculation. Therefore both steps should in a consequent manner be practised upon symmetry. In section 2.4. this was entirely on an intuitive level done for the multi-centre integrals. Here, it is a bit more involved, and the group-theoretical treatment of the symmetry properties of LCAO wavefunctions is given in short in the first section of this chapter. Then, in 5.2., a decomposition of the LCAO electron density into site densities is introduced. As rather multiform spatial dependecies have to be treated, again a local basis function expansion is most useful. The symmetry reduction of the resulting expressions is performed in section 5.3. As described in the last section of this chapter, the resulting expressions are composed of three contributions of increasing complexity: the atom core density contribution, the atomic net density contribution expressed by one-centre matrix elements, and the overlap density contribution expressed in terms of two-centre matrix elements. The conversion of the results into a useful basis function representation of the total density within an atom volume instead of the (overlapping) site densities is also given there.

5.1. Symmetry

Apart from Bloch's theorem and some simple considerations of the spatial inversion in section 2.3., and from the rotation of spherical harmonics given in Appendix 1., we have not up to now made explicit use of symmetry considerations. The computation and diagonalization of the LCAO matrices is so simple that there is no need in a symmetry reduction except for the business of local and global coordinates explained in section 2.4. and in Appendix 1. The symmetry classification of the states (2.51) can in most cases be made intuitively just by inspection of the coefficients $C_{Sn;\nu}$ of atom-like angular momentum eigenstates n at atom sites **S**. As the calculation of the total electronic charge density from the LCAO representation implies a larger computational effort, a little more advanced symmetry considerations may be useful here. True, in the case of a Bravais lattice everything can again easily be done on an intuitive level. But our treatment aims at a more general case. The details of what is sketched in the following may be found, e.g., in [66].

The group of translations of a crystal lattice is

$$\mathcal{T} = \{\ \hat{T}_{\mathbf{R}}\ |\ \mathbf{R}\ \text{being any lattice vector}\ \} \tag{5.1}$$

The space group $\mathcal{G}$ as the full symmetry group of the crystal structure contains besides the elements $\hat{T}_{\mathbf{R}}$ additional transformations of the general type

$$\hat{U} = \hat{T}_{\mathbf{U}}\hat{u} = \hat{u}\hat{T}_{\mathbf{U}}\ ,\qquad \hat{u}\mathbf{u} = \mathbf{u}\ , \tag{5.2}$$

where **U** is different from every non-zero lattice vector (actually it is either a certain fraction of a lattice vector or, in most cases, **U** = 0), and $\hat{u}$ is a homogeneous linear transformation, a proper or improper rotation or a reflection, of the vectors **r** − **u** leaving both the centre **u** of the transformation and the glide vector **U** invariant. In full detail,

$$\hat{u} = \hat{T}_{\mathbf{u}}\hat{u}_o\hat{T}_{-\mathbf{u}}\ ,\qquad \hat{U} = \hat{T}_{\mathbf{u}}\hat{u}_o\hat{T}_{-\mathbf{v}}\ ,\qquad \mathbf{U} = \mathbf{u} - \mathbf{v}\ , \tag{5.3}$$

where $\hat{u}_o$ is a homogeneous transformation leaving the origin of the coordinate system invariant, and $\hat{u}_o\mathbf{U} = \mathbf{U}$. That $\hat{U}$ is a symmetry transformation of the crystal lattice implies particularly that for any pair of translationally equivalent lattice points the transformed points are again translationally equivalent. Since translations do not change distance vectors between two points, it follows that $\mathbf{R}' = \hat{u}_o\mathbf{R}$ is a lattice vector for any lattice vector **R**. Hence, for any **R**,

$$\hat{T}_{\mathbf{R}}\hat{U} = \hat{T}_{\mathbf{R}}\hat{T}_{\mathbf{U}}\hat{u} = \hat{T}_{\mathbf{U}}\hat{T}_{\mathbf{R}}\hat{u} = \hat{T}_{\mathbf{U}}\hat{u}\hat{T}_{\hat{u}_o^{-1}\mathbf{R}} = \hat{U}\hat{T}_{\mathbf{R}'}\ . \tag{5.4}$$

In other words, $\mathcal{T}$ is an invariant subgroup of the space group $\mathcal{G}$.

Let $\hat{U} \neq \hat{U}'$ be two elements of the space group, for which $\hat{u} = \hat{u}'$. Then $\hat{U}^{-1}\hat{U}' = \hat{T}_{\mathbf{U}'-\mathbf{U}}$ is a translation, and hence **U'**−**U** must be a lattice vector, i.e., $\hat{U}' = \hat{U}\hat{T}_{\mathbf{R}}$. We see that up to lattice translations there is a one-to-one correspondence between the point transformations $\hat{u}$ and the glide vectors **U** in (5.2). Let now $\hat{u}_o = \hat{u}'_o$, but $\mathbf{u} \neq \mathbf{u}'$, i.e., the two space group operations refer to two different centres **u** and **u'** but are otherwise the same. Then (note that $\hat{u}$ and $\hat{u}^{-1}$ have the same centre **u**)

$$\hat{U}^{-1}\hat{U}' = \hat{T}_{-\mathbf{U}}\hat{T}_{\mathbf{u}}\hat{u}_o^{-1}\hat{T}_{-\mathbf{u}}\hat{T}_{\mathbf{u}'}\hat{u}_o\hat{T}_{-\mathbf{u}'}\hat{T}_{\mathbf{U}} = \hat{T}_{\hat{u}_o^{-1}(\mathbf{u}'-\mathbf{u}) - (\mathbf{u}'-\mathbf{u})} \tag{5.5}$$

is again a translation, and hence $\mathbf{R} = \hat{u}_o^{-1}(\mathbf{u}'-\mathbf{u})-(\mathbf{u}'-\mathbf{u})$ a lattice vector. Furthermore, slightly rewriting (5.3), for this R,

$$\hat{U}' = \hat{u}_o\hat{T}_{\mathbf{U}+\hat{u}_o^{-1}\mathbf{u}'-\mathbf{u}'} = \hat{u}_o\hat{T}_{\mathbf{U}+\hat{u}_o^{-1}\mathbf{u}-\mathbf{u}}\hat{T}_{\mathbf{R}} = \hat{U}\hat{T}_{\mathbf{R}} , \tag{5.6}$$

that is, again there is up to lattice translations a one-to-one correspondence between the point transformations $\hat{u}_o$ and their centres **u**. (The special case of this situation with $\hat{u}_o$ being the spatial inversion $\hat{I}$ was considered in section 2.3.)

What we have proved by these considerations is a true one-to-one correspondence between the different cosets $\hat{U}\mathcal{T}$ in the space group $\mathcal{G}$ with respect to the group of translations $\mathcal{T}$ and the different point transformations $\hat{u}_o$ contained in the elements of $\mathcal{G}$. Since $\mathcal{T}$ was shown to be an invariant subgroup of $\mathcal{G}$, the cosets $\hat{U}\mathcal{T}$ form the factorgroup $\mathcal{g} = \mathcal{G}/\mathcal{T}$. As is easily seen from these considerations, the correspondence of the cosets $\hat{U}\mathcal{T}$ with the point transformations $\hat{u}_o$ is in fact an isomorphism of groups so that the set of all point transformations $\hat{u}_o$ contained in the elements of $\mathcal{G}$ forms also a group, the point group of the crystal lattice, isomorphic to $\mathcal{g}$ and therefore also denoted by $\mathcal{g}$.

Contrary to this, the elements $\hat{U}$ of (5.2) and (5.3) do in general not form a group: products of them may result in pure translations out of $\mathcal{T}$. If, however, all transformations $\hat{u}_o$ themselves are symmetry transformations of the crystal lattice, then $\mathcal{g}$ is an invariant subgroup of $\mathcal{G}$, and $\mathcal{G}$ is a direct product of groups:

$$\mathcal{G} = \mathcal{g} \times \mathcal{T}, \quad \mathcal{T} = \mathcal{G}/\mathcal{g}, \quad \mathcal{g} = \mathcal{G}/\mathcal{T}. \tag{5.7}$$

In this case $\mathcal{G}$ is called symmorphic, otherwise non-symmorphic. Examples of symmorphic groups are the space groups of all Bravais lattices. Examples of non-symmorphic groups are the space groups of the h.c.p. and diamond lattices.

The performance of the symmetry operations for the various physical quantities as the atom sites **R+S**, the Bloch states $|\mathbf{k}\nu\rangle$, the vibrational modes, and so on results in the various representations of the symmetry groups. For the sake of simplicity of notation we shall use in all these cases the same notation. So, if $\hat{U}\hat{T}_{\mathbf{R}}$ is an element of $\mathcal{G}$, then

$$\mathbf{R}'+\mathbf{S}' = \hat{U}\hat{T}_{\mathbf{R}}(\mathbf{R}+\mathbf{S}) \tag{5.8}$$

denotes the corresponding transformation of atom sites,

$$|\mathbf{k}\nu>' = \hat{U}\hat{T}_{\mathbf{R}}|\mathbf{k}\nu> \tag{5.9}$$

that of electron states, and so on. All the transformations will be understood in the active sense here.

The crystal Hamiltonian $\hat{h}$ commutes with all elements of $\mathcal{G}$ (in the sense of (5.9)). Hence, if $|\mathbf{k}\nu>$ is a solution of the Schrödinger equation (2.3) for a crystal, then

$$\hat{U}\hat{T}_{\mathbf{R}}|\mathbf{k}\nu>\ \varepsilon_{\nu}(\mathbf{k}) = \hat{U}\hat{T}_{\mathbf{R}}\hat{h}|\mathbf{k}\nu> = \hat{h}\hat{U}\hat{T}_{\mathbf{R}}|\mathbf{k}\nu>\ , \tag{5.10}$$

that is, $\hat{U}\hat{T}_{\mathbf{R}}|\mathbf{k}\nu>$ is again a solution of (2.3) for the same energy, of course not necessarily linearly independent from the previous one. In the well known manner the states $|\mathbf{k}\nu>$ may be chosen to form bases of irreducible representations of the group $\mathcal{G}$. From (2.48) we have

$$\hat{T}_{\mathbf{R}}|\mathbf{k}\nu> = |\mathbf{k}\nu>\ e^{i\mathbf{k}\mathbf{R}}, \tag{5.11}$$

and together with (5.4)

$$\hat{T}_{\mathbf{R}}\hat{U}|\mathbf{k}\nu> = \hat{U}\hat{T}_{\hat{u}_o^{-1}\mathbf{R}}|\mathbf{k}\nu> = \hat{U}|\mathbf{k}\nu>\ e^{i\mathbf{k}\hat{u}_o^{-1}\mathbf{R}} = \hat{U}|\mathbf{k}\nu>\ e^{i\hat{u}_o\mathbf{k}\mathbf{R}}, \tag{5.12}$$

meaning that $\hat{U}|\mathbf{k}\nu>$ is a state $|\hat{u}_o\mathbf{k}\nu'>$ belonging to the wave-vector $\hat{u}_o\mathbf{k}$ with

$$\varepsilon_{\nu'}(\hat{u}_o\mathbf{k}) = \varepsilon_{\nu}(\mathbf{k}), \tag{5.13}$$

that is, in general,

$$\hat{U}|\mathbf{k}\nu> = \sum_{\nu'} |\hat{u}_o\mathbf{k}\nu'>\ U_{\nu'\nu}\ , \tag{5.14}$$

where the summation is over all states ν' energetically degenerate according to (5.13). Hence, the space group $\tilde{\mathcal{G}}$ of the reciprocal lattice of quasi-momenta $\mathbf{k}$ of any crystal lattice is always a symmorphic group

$$\tilde{\mathcal{G}} = g \times \tilde{\mathcal{T}}, \tag{5.15}$$

independent on whether $\mathcal{G}$ was symmorphic, $\mathcal{G} = g \times \mathcal{T}$, or not.

Since Schrödinger's equation is invariant under time inversion meaning for the wavefunction complex conjugation and hence, from (2.48), $\mathbf{k} \to -\mathbf{k}$ as it should be from the physical meaning of $\mathbf{k}$,

$$\varepsilon_\nu(\mathbf{k}) = \varepsilon_\nu(-\mathbf{k}) \tag{5.16}$$

always holds, eventually additionally to (5.13) and independently on whether the point group $\mathcal{G}$ contains the spatial inversion or not. However, (5.16) causes no additional symmetry for the spatial dependence of the wavefunctions. (If there would be an external magnetic field or a spontaneous magnetization, the magnetic field would have to be reversed in (5.16) simultaneously with the wave-vector.)

Now we consider our LCAO representation (2.50-51) of the Bloch states $|\mathbf{k}\nu\rangle$, that is,

$$|\mathbf{k}\nu\rangle = \frac{1}{\sqrt{N}} \sum_{\mathbf{RS}n} |\mathbf{R}+\mathbf{S}n\rangle \, e^{i\mathbf{k}(\mathbf{R}+\mathbf{S})} \, C_{\mathbf{S}n;\nu}(\mathbf{k}) \, . \tag{5.17}$$

Operating on it with $\hat{U}$ yields

$$\hat{U}|\mathbf{k}\nu\rangle = \frac{1}{\sqrt{N}} \sum_{\mathbf{RS}n} \sum_{n'} |\hat{U}(\mathbf{R}+\mathbf{S})n'\rangle \, u_{n'n} \, e^{i\mathbf{k}(\mathbf{R}+\mathbf{S})} \, C_{\mathbf{S}n;\nu}(\mathbf{k}) \, , \tag{5.18}$$

where the matrix $u_{n'n}$ is again that of rotating the spherical harmonics as given in Appendix 1. Relabelling the summation variables this last expression may be rewritten as

$$\hat{U}|\mathbf{k}\nu\rangle = \frac{1}{\sqrt{N}} \sum_{\mathbf{RS}n} \sum_{n'} |\mathbf{R}+\mathbf{S}n'\rangle \, u_{n'n} \, e^{i\mathbf{k}\hat{U}^{-1}(\mathbf{R}+\mathbf{S})} \, C_{\hat{U}^{-1}\mathbf{S}n;\nu}(\mathbf{k}) \, . \tag{5.19}$$

On the other hand, (5.14) asserts that this expression is equal to

$$\sum_{\nu'} |\hat{u}_o\mathbf{k}\nu'\rangle \, U_{\nu'\nu} = \sum_{\nu'} \frac{1}{\sqrt{N}} \sum_{\mathbf{RS}n} |\mathbf{R}+\mathbf{S}n\rangle \, e^{i\mathbf{k}\hat{u}_o^{-1}(\mathbf{R}+\mathbf{S})} \, C_{\mathbf{S}n;\nu'}(\hat{u}_o\mathbf{k}) \, U_{\nu'\nu} \tag{5.20}$$

From (5.2) and (5.3),

$$\hat{U}^{-1} = \hat{T}_{-\mathbf{U}+\mathbf{u}-\hat{u}_o\mathbf{u}} \, \hat{u}_o^{-1} \, . \tag{5.21}$$

Comparing (5.19) with (5.20) yields the general symmetry property of the LCAO coefficients:

$$\boxed{C_{\mathbf{S}n;\nu}(\hat{u}_o\mathbf{k}) = e^{-i\mathbf{k}(\mathbf{U}-\mathbf{u}+\hat{u}_o\mathbf{u})} \sum_{\nu'} \sum_{n'} u_{nn'}\, C_{\hat{U}^{-1}\mathbf{S}n';\nu'}(\mathbf{k})\, U^{-1}_{\nu'\nu}} \tag{5.22}$$

Let M be the order of the point group $\mathcal{G}$, that is the number of different elements $\hat{u}_o$. For a general wave-vector $\mathbf{k}$ there may be M different wave-vectors $\hat{u}_o\mathbf{k}$ forming the so-called Wigner star or the orbit of the vector $\mathbf{k}$. For such $\mathbf{k}$ there is no symmetry degeneracy of energies $\varepsilon_\nu(\mathbf{k})$ with $\mathbf{k}$ fixed, and the sum over ν' in (5.14) and (5.22) is to be omitted the matrix $U_{\nu'\nu}$ being replaced by unity. If $\mathbf{k}$ points into a symmetry direction, then the number of physically different wave-vectors in the Wigner star is only a fraction $M_\mathbf{k}$ of M, whereas $M/M_\mathbf{k}$ transformations $\hat{u}_o \in \mathcal{G}$ leave the wave-vector $\mathbf{k}$ invariant up to a reciprocal lattice translation. Clearly these elements again form a group, the subgroup $\mathcal{G}_\mathbf{k}$ of $\mathcal{G}$. If $\hat{w}_o \in \mathcal{G}_\mathbf{k}$, that is, $\hat{w}_o\mathbf{k} = \mathbf{k} \bmod \tilde{\mathcal{T}}$, then $\hat{w}'_o = \hat{u}_o\hat{w}_o\hat{u}_o^{-1}$ leaves $\hat{u}_o\mathbf{k} \bmod \tilde{\mathcal{T}}$ invariant for any $\hat{u}_o \in \mathcal{G}$, i.e., $\hat{w}'_o \in \mathcal{G}_{\hat{u}_o\mathbf{k}}$, and vice versa. On the other hand, the mapping

$$\mathcal{G}_\mathbf{k} \rightarrow \mathcal{G}_{\hat{u}_o\mathbf{k}} : \hat{w}_o \rightarrow \hat{u}_o\hat{w}_o\hat{u}_o^{-1} \tag{5.23}$$

is an isomorphism of groups, and consequently all groups $\mathcal{G}_{\hat{u}_o\mathbf{k}}$, $\hat{u}_o \in \mathcal{G}$ are isomorphic, which was to be expected because all wave-vectors $\hat{u}_o\mathbf{k}$ of a Wigner star are symmetry equivalent.

Let first the space group $\mathcal{G}$ of the crystal lattice be symmorphic. In this case all Bloch states of the crystal may be chosen to form bases of irreducible representations of the groups $\mathcal{G}_\mathbf{k}$, and if $\hat{u}_o \in \mathcal{G}_\mathbf{k}$, then $U_{\nu'\nu}$ of (5.14) and (5.22) is the corresponding representation matrix of the group element $\hat{u}_o$. In general, if $\hat{u}_o$ in (5.14) or (5.22) runs through the full point group $\mathcal{G}$ and $\mathbf{k}$ runs through its Wigner star, these equations may be read as a (in general reducible) representation of $\mathcal{G}$ whose restriction on $\mathcal{G}_\mathbf{k}$ and on the state space spanned by the $|\mathbf{k}\nu\rangle$ with $\mathbf{k}$ fixed is the above considered irreducible representation of $\mathcal{G}_\mathbf{k}$. The isomorphic mappings (5.23) ensure that the same is true for the restrictions with respect to any other member $\hat{u}_o\mathbf{k}$ of the Wigner star of $\mathbf{k}$, and that all these restrictions form unitary equivalent irreducible representations of the isomorphic to each other groups $\mathcal{G}_{\hat{u}_o\mathbf{k}}$.

In the case of a non-symmorphic space group there may be elements $\hat{u}_o \in \mathcal{G}_\mathbf{k}$ for which a fixed non-unit phase factor $\exp[-i\mathbf{k}(\mathbf{U}-\mathbf{u}+\hat{u}_o\mathbf{u})]$ appears in the representations transforming Bloch wavefunctions as is seen from (5.22). The thereby appearing additional complication is connected with the fact that even for a product $\hat{u}_o . \hat{u}'_o ...$ of elements of $\mathcal{G}_\mathbf{k}$ resulting in the unit element the corresponding product of phase factors need not be unity. Such generalized representations are called multiplier representations or projective representations. For instance the space group of the h.c.p. lattice contains a screw displacement $\hat{U}$ consisting of a 180° rotation $\hat{u}_o$ around the c-axis and a simultaneous displacement $\mathbf{U} = c/2\ \mathbf{e}_c$. Besides, the spatial inversion $\hat{I}_o$ at the origin is also an element of this space group. Thus $\hat{u}_o \hat{I}_o$ is an element of the group $\mathcal{G}_\mathbf{k}$ for any wave-vector $\mathbf{k}$ with $k_c = \pi/c$, that is on the top plane of the Brillouin zone. The corresponding phase factor is $\exp(-i\mathbf{kU}) = -i$. Although $(\hat{u}_o \hat{I}_o)^2 = 1$, this square element is connected with a phase factor -1. There is, however, always a small integer m so that the m-th power of these strange phase factors is unity (in the given example m=2). The multiplier representations may therefore always be obtained from ordinary representations of a manifolded group $\tilde{\mathcal{G}}_\mathbf{k}$. In the given example $\tilde{\mathcal{G}}_\mathbf{k}$ is obtained by distinguishing $(\hat{u}_o \hat{I}_o)^2$ from 1 and only putting $(\hat{u}_o \hat{I}_o)^{2m} = 1$. In the h.c.p. structure the consequence is a twofold degeneracy of all energy bands on the top plane of the Brillouin zone as only two-dimensional irreducible representations of $\tilde{\mathcal{G}}_\mathbf{k}$ appear for which $(\hat{u}_o \hat{I}_o)^2$ is represented by -1. It is generally the group $\tilde{\mathcal{G}}_\mathbf{k}$ (coniciding with $\mathcal{G}_\mathbf{k}$ in symmorphic structures) which is called the wave-vector group of the wave-vector $\mathbf{k}$.

5.2. The total electron density

In the sense of the density functional theory of section 1.2., the total electron density of the electronic ground-state of the system is obtained according to (1.20) by summing up over the squares of all occupied one-electron wavefunctions,

$$\varrho(\mathbf{r}) = \sum_{\nu} n(\varepsilon_\nu)\ (\mathbf{r}|\nu)(\nu|\mathbf{r}) \ , \tag{5.24}$$

where

$$n(\varepsilon) = \Theta(\varepsilon_F - \varepsilon) = \begin{cases} 1 \\ 0 \end{cases} \quad \text{for} \quad \varepsilon \lessgtr \varepsilon_F \tag{5.25}$$

is the Fermi function for T=0. With the LCAO representation (2.29) this is given by a double sum over the atom sites

$$\varrho(\mathbf{r}) = \sum_{\nu} n(\varepsilon_\nu) \sum_{\mathbf{R}'m} \sum_{\mathbf{R}n} \langle\mathbf{r}|\mathbf{R}'m\rangle\ c_{\mathbf{R}'m;\nu} c^*_{\mathbf{R}n;\nu}\ \langle\mathbf{R}n|\mathbf{r}\rangle \tag{5.26}$$

which may in a natural way be understood as a sum over site densities $\varrho_\mathbf{R}$:

$$\varrho(\mathbf{r}) = \sum_{\mathbf{R}} \varrho_{\mathbf{R}}(\mathbf{r}-\mathbf{R}) ,$$

$$\varrho_{\mathbf{R}}(\mathbf{r}-\mathbf{R}) = \sum_{\nu} n(\varepsilon_\nu) \sum_{\mathbf{R}'} \sum_{mn} \langle \mathbf{r}|\mathbf{R}'m\rangle \, c_{\mathbf{R}'m;\nu} \, c^*_{\mathbf{R}n;\nu} \, \langle \mathbf{R}n|\mathbf{r}\rangle . \tag{5.27}$$

The remaining lattice sum over $\mathbf{R}'$ in the last expression for $\varrho_{\mathbf{R}}$ converges exponentially due to the exponential decay of the product $\langle \mathbf{r}|\mathbf{R}'m\rangle\langle \mathbf{R}n|\mathbf{r}\rangle$ of the two overlapping wavefunctions centred at $\mathbf{R}$ and $\mathbf{R}'$. The remaining part of this chapter is concerned with the calculation of $\varrho_{\mathbf{R}}$.

Of course, depending on whether the ν-sum of (5.27) runs only over valence states or over all states including ionic core states, the valence electron density or the total electron density is obtained. Both cases have their justified applications; we shall, however, everywhere in the following consider the total electron density. In any case, (5.27) contains via (2.38) core-orthogonalization contributions to the densities. According to (2.38) these corrections contain a small factor, $(\mathbf{R}''c|\mathbf{R}n) < 0.02$ for our optimum basis functions. There are first order corrections in (5.27) where $\mathbf{R}'' = \mathbf{R}'$ is the centre of the other valence function, and second order corrections where $\mathbf{R}'' \neq \mathbf{R}'$ is a third centre so that another small factor of neighbouring core-valence overlap is present. Despite the smallness of the latter they are considerable in number so that their total contribution to $\varrho_{\mathbf{R}}$ amounts up to one tenth of an electron. The integral contribution of the first order corrections is exactly zero as core and valence wavefunctions at the same centre are orthogonal to each other. So these corrections result in a displacement of about one hundredth of an electron by a few tenth of a Bohr radius within the ionic core. At least as the effect on the self-consistent potential is considered these corrections may be neglected. Hence, we are generally left with non-degenerate three-centre corrections which may be incorporated below in (5.31) in the spirit of (2.77). We rewrite the site density as

$$\varrho_{\mathbf{R}}(\mathbf{r}) = \sum_{\nu} n(\varepsilon_\nu) \sum_{\mathbf{R}'} \sum_{mn} \langle \mathbf{r}|\mathbf{R}'-\mathbf{R}m\rangle \, c_{\mathbf{R}'m;\nu} \, c^*_{\mathbf{R}n;\nu} \, \langle 0n|\mathbf{r}\rangle , \tag{5.28}$$

where $\langle \mathbf{r}+\mathbf{R}|\mathbf{R}'m\rangle = \langle \mathbf{r}|\mathbf{R}'-\mathbf{R}m\rangle$ was used.

Instead of calculating (5.28) for a suitable mesh of $\mathbf{r}$-points it is convenient again to apply the machinery of basis function expansions

$$\boxed{\varrho_{\mathbf{R}}(\mathbf{r}) = \sum_{t} \langle \mathbf{r}|t\rangle \, \varrho_t^{(\mathbf{R})}} \tag{5.29}$$

as in section 2.1. Normalizing the basis states $|t\rangle$ (again preferably of Slater type) as in (2.16), we find in analogy to (2.19) for the coefficients

$$\boxed{\varrho_t^{(R)} = \sum_{t'} \langle \varrho_R | t' \rangle_w \, n_{t'}^w \, (D^{-1} D^{-1T})_{t't} \, n_t^w ,} \tag{5.30}$$

where the **D**-matrix is obtained from (2.18), and only the matrix elements $\langle \varrho_R | t \rangle_w$ are left to be calculated from (5.28):

$$\langle \varrho_R | t \rangle_w = \sum_{\nu} n(\varepsilon_\nu) \sum_{R'} \sum_{mn} c^*_{Rn;\nu} \langle Rn | t_R | R'm \rangle_w \, c_{R'm;\nu} \tag{5.31}$$

where we have attached with t an index **R** indicating the site at which the basis function t is centred. The matrix element in this expression means a three-dimensional integral with one atom-like wavefunction and the basis function $b_t(\mathbf{r})$ centred at **R**, and a second atom-like wavefunction centred at **R'**. Eventually the integrand may contain an additional weighting factor $w(\mathbf{r})$ for which our experience says $w(\mathbf{r}) = r^2$ (in addition to the Jacobian of spherical coordinates) being very cooperative. This matrix element may be expressed in terms of two-centre integrals of the type considered in section 2.4. (formulae (2.66-72)).

5.3. The case of a crystal lattice

If the atom positions **R** form a crystal lattice, in (5.24) the states $|\nu\rangle$ are to be replaced by the states (5.17) with the result

$$\boxed{\varrho(\mathbf{r}) = \sum_{\mathbf{RS}} \varrho_{\mathbf{S}}(\mathbf{r}-\mathbf{R}-\mathbf{S})} \tag{5.32}$$

instead of (5.27). Due to the translational invariance of the crystal lattice all site densities centred at lattice sites **R+S** with different **R** but the same **S** are identical. Using the representation (5.29), the coefficients $\varrho_t^{(S)}$ are obtained from the same formula (5.30) only with (5.31) replaced by

$$\langle \varrho_{\mathbf{S}} | t \rangle_w = \frac{\mathrm{Re}}{N} \sum_{\mathbf{k}\nu} n(\varepsilon_\nu(\mathbf{k})) \sum_{\mathbf{S}'} \sum_{mn} c^*_{\mathbf{S}n;\nu}(\mathbf{k}) \; *$$

$$* \left[\sum_{\mathbf{R}} e^{i\mathbf{k}(\mathbf{R}+\mathbf{S}'-\mathbf{S})} \langle \mathbf{S}n | t_{\mathbf{S}} | \mathbf{R}+\mathbf{S}'m \rangle_w \right] c_{\mathbf{S}'m;\nu}(\mathbf{k}) \; . \tag{5.33}$$

Since $\varepsilon_\nu(\mathbf{k}) = \varepsilon_\nu(-\mathbf{k})$ always holds according to (5.17) and $\varphi_{\mathbf{k}\nu}(\mathbf{r}) = \varphi^*_{-\mathbf{k}\nu}(\mathbf{r})$, the site density $\varrho_{\mathbf{S}}(\mathbf{r})$ defined by (5.27) is indeed real. (The total density $\varrho(\mathbf{r})$ is real by the definition (5.24). The spin-polarized case in the presence of an external or internal magnetic field, where the analogous site contributions to the diagonal elements of the spin density matrix need not necessarily be real, is not treated here. If spin-orbit coupling is neglected, the reality of the site densities is preserved.) Therefore we were allowed to introduce the sign of the real part in (5.33) without harm thus reducing the numerical effort by a factor of two.

Standard methods of self-consistent electronic structure calculations use in most cases a muffin-tin approximation for the charge density: a spherically symmetric charge density within touching each other muffin-tin spheres centred at the atom positions (and eventually at additional interstitial positions) and a constant charge density in the interstitial region inbetween the muffin-tin spheres. Although LCAO methods are not at all bound to muffin-tin constructions the numerics is very much simplified if the site densities may be spherically averaged giving then rise to spherically symmetric site potentials. There is, however, no restriction with respect to the overlap of neighbouring site densities and site contributions to the Coulomb potential, respectively. In metallic systems having more or less close-packed structures there is no problem with the spherical approximation of $\varrho_{\mathbf{S}}(\mathbf{r})$. In the case of open structures the most convenient way is to introduce interstitial sites as discussed in section 3.2. for the case of the diamond lattice. The interstitial sites may then have their own site densities and site potentials. The spherical approximation is then expected again not to cause any noticable inaccuracy. The spherical average of $\varrho_{\mathbf{S}}(\mathbf{r})$ is simply obtained by using spherically symmetric basis functions $\langle \mathbf{r}|t\rangle = \langle r|t\rangle$ in (5.29) and (5.33), preferably

$$b_t(r) = \langle r|t\rangle = r^{n_t-1}\, e^{-\alpha_t r} . \qquad (5.34)$$

Approved parameters are $n_t = 1,2,3$; $\alpha_t = 2A, 2A/q, \ldots, 4.4/r_{at}$ with one more α-value than in the wavefunction basis (A again being the atomic number). This spherical approximation immediately allows for a couple of further simplifications.

Consider the contribution with a fixed wave-vector $\mathbf{k}$ to (5.33) and apply an element of the point group $\mathcal{G}$ of the crystal to that wave-vector: $\mathbf{k}' = \hat{u}_o\mathbf{k}$. First of all the spherically symmetric function b_t is invariant under the rotation and/or reflection $\hat{u}_o$: $\hat{u}_o^+ t \hat{u}_o = t$ in (5.33). Consequently, as the application of the corresponding space group element $\hat{U}$ of (5.2-3) does not change distances, we have for some $\mathbf{R}'$

$$\langle \mathbf{S}n|\hat{U}\, t_{\hat{U}^{-1}\mathbf{S}}\, \hat{U}^{-1}|\mathbf{R}'+\mathbf{S}'m\rangle_w = \sum_{m'n'} u^{-1}_{nn'} \langle \mathbf{S}n'|t_{\mathbf{S}}|\mathbf{R}+\mathbf{S}'m'\rangle_w\, u^{-1}_{m'm} \qquad (5.35)$$

where the transformation matrices $\mathbf{u}$ are the same as in (5.19). Along the same lines,

$$e^{i(\hat{u}_o\mathbf{k})(\mathbf{R}+\mathbf{S}'-\mathbf{S})} = e^{i\mathbf{k}\hat{U}^{-1}(\mathbf{R}'+\mathbf{S}'-\mathbf{S})}. \tag{5.36}$$

In a non-Bravais lattice $\hat{U}\mathbf{S}$ is not necessarily equal to $\mathbf{S}$, but for symmetry reasons the site densities $\varrho_\mathbf{S}(\mathbf{r})$ and $\varrho_{\hat{U}\mathbf{S}}(\mathbf{r})$ can only differ from each other by a rotation $\hat{u}_o$ and hence their spherical averages must be equal. Therefore, understanding b_t in (5.33) as a spherically symmetric function (5.34), we may average the expression (5.33) over all symmetry equivalent sites $\hat{U}\mathbf{S}$ their number being $n_\mathbf{S}$:

$$\langle\varrho_\mathbf{S}|t\rangle_W = \frac{\mathrm{Re}}{Nn_\mathbf{S}} \sum_{\mathbf{k}\nu} n(\varepsilon_\nu(\mathbf{k})) \sum_{\mathbf{S}''n}^{\{\hat{U}\mathbf{S}\}} \sum_{\mathbf{S}'m} C^*_{\mathbf{S}''n;\nu}(\mathbf{k}) \, * \\ * \left[\sum_{\mathbf{R}} e^{i\mathbf{k}(\mathbf{R}+\mathbf{S}'-\mathbf{S}'')} \langle\mathbf{S}''n|t_{\mathbf{S}''}|\mathbf{R}+\mathbf{S}'m\rangle_W\right] C_{\mathbf{S}'m;\nu}(\mathbf{k}) \,. \tag{5.37}$$

Note that the expression in square brackets has the same structure as the LCAO matrices (2.53) and (2.55), and its computation may be organized in the same way.

As in section 2.3. profit can be derived from combining time inversion (complex conjugation) with the spatial inversional symmetry. If we denote the square bracket expression of (5.37) by $T^t_{\mathbf{S}''n;\mathbf{S}'m}(\mathbf{k})$, then, because the summation over $\mathbf{S}''$ always contains $\hat{I}\mathbf{S}$ together with $\mathbf{S}$, $\sum\sum \mathrm{Re}[\mathbf{C}^*\mathbf{T}^t\mathbf{C}] = \sum\sum \tilde{\mathbf{C}}^T \mathrm{Re}[\mathbf{U}^+\mathbf{T}^t\mathbf{U}]\, \tilde{\mathbf{C}}$, where $\mathbf{U}$ means the unitary transformation (2.61) and $\tilde{\mathbf{C}}$ are the real wavefunction coefficients obtained from the eigenvalue problem with the real matrices $\tilde{\mathbf{H}}$ and $\tilde{\mathbf{B}}$ of (2.62). Note, however, that $\tilde{\mathbf{T}}^t = \mathbf{U}^+\mathbf{T}^t\mathbf{U}$ has not the simplified structure (2.62), because, in general, $\mathbf{T}^t$ is not Hermitean due to the linkage of $t_\mathbf{S}$ with the left index of the matrix $\mathbf{T}^t$. However, instead of $\mathbf{T}^t$ the two matrices

$$T^{\pm t}_{\mathbf{S}n;\mathbf{S}'m}(\mathbf{k}) = \sum_{\mathbf{R}} e^{i\mathbf{k}(\mathbf{R}+\mathbf{S}'-\mathbf{S})} \langle\mathbf{S}n|t_\mathbf{S} \pm t_{\mathbf{R}+\mathbf{S}'}|\mathbf{R}+\mathbf{S}'m\rangle$$

may be used where both matrices $\mathbf{T}^{\pm t}$ have the property analogous to (2.57) $\mathbf{T}^{+t}$ being Hermitean and $\mathbf{T}^{-t}$ being anti-Hermitean. Both their transforms $\tilde{\mathbf{T}}^{+t}$ and $\tilde{\mathbf{T}}^{-t}$ are real, $\tilde{\mathbf{T}}^{+t}$ having the structure of (2.62) and $\tilde{\mathbf{T}}^{-t}$ having a structure which only differs from (2.62) by a minus sign at the blocks with $(\tilde{\mathbf{T}}^{-t})^T_{IA}$. As in the Hermitean (real symmetric) case of the matrices $\mathbf{H}$ ($\tilde{\mathbf{H}}$) and $\mathbf{B}$ ($\tilde{\mathbf{B}}$), it again suffices for $\mathbf{T}^{\pm t}$ ($\tilde{\mathbf{T}}^{\pm t}$) to calculate one triangle and then to complete the matrices using (anti-)Hermicity (real (anti-)symmetry). Thus, for the computation of **H**, **B**, and $\mathbf{T}^{\pm t}$ ($\tilde{\mathbf{H}}$, $\tilde{\mathbf{B}}$, and $\tilde{\mathbf{T}}^{\pm t}$), the same computer code may be used. Finally, $\mathbf{T}^{t} = (\mathbf{T}^{+t} + \mathbf{T}^{-t})/2$ ($\tilde{\mathbf{T}}^{t} = (\tilde{\mathbf{T}}^{+t} + \tilde{\mathbf{T}}^{-t})/2$).

Although the additional summation over **S**" in (5.37) seems to be a strange procedure it causes a very large gain. Namely, if we now consider the contribution of the wave-vector $\hat{u}_o\mathbf{k}$ to (5.37) and use (5.22) as well as (5.35-36), we find

$$\sum_{\mathbf{S}''n}^{\{\hat{U}\mathbf{S}\}} \sum_{\mathbf{S}'m} \sum_{\nu'\nu''} U^{-1*}_{\nu''\nu} \, C^{*}_{\hat{U}^{-1}\mathbf{S}''n;\nu''}(\mathbf{k}) \; *$$

$$* \sum_{\mathbf{R}} e^{i\mathbf{k}\hat{U}^{-1}(\mathbf{R}'+\mathbf{S}'-\mathbf{S}'')} \langle \mathbf{S}''n | \hat{U} \, t_{\hat{U}^{-1}\mathbf{S}''} \, \hat{U}^{-1} | \mathbf{R}'+\mathbf{S}'m \rangle_W \; *$$

$$* \, C_{\hat{U}^{-1}\mathbf{S}'m;\nu'}(\mathbf{k}) \, U^{-1}_{\nu'\nu} \tag{5.38}$$

where the matrices **u** of (5.35) have cancelled against those of (5.22). According to (5.15), the summation over ν' and ν'' is over all states energetically degenerate with the state ν, and the matrix **U** is unitary. (5.37) contains a summation over the degenerate ν which results after inserting (5.38) in

$$\sum_{\nu} U^{-1*}_{\nu''\nu} \, U^{-1}_{\nu'\nu} = \sum_{\nu} U^{-1}_{\nu'\nu} \, U_{\nu\nu''} = \delta_{\nu'\nu''} \; . \tag{5.39}$$

Relabelling the summation variables $\mathbf{R} \rightarrow \hat{U}\mathbf{R}'$, $\mathbf{S}' \rightarrow \hat{U}\mathbf{S}'$, and $\mathbf{S}'' \rightarrow \hat{U}\mathbf{S}''$ we readily find that the contribution of $\hat{u}_o\mathbf{k}$ to (5.37) is exactly equal to that of **k**, so that the **k**-dependend expression of (5.37) has the full point symmetry of the Brillouin zone, and the **k**-integration may be performed over the irreducible part of the Brillouin zone only. Thus, e.g. for the diamond lattice, by introducting an additional summation over the two sites in the unit cell in replacing (5.33) by (5.37) the integration range in the Brillouin zone is reduced by a factor of 48. The problem of **k**-space integrations is treated in the next chapter.

5.4. Core, net, and overlap densities

The basic contributions to (5.37) and (5.31), respectively, are those containing the one-centre matrix elements and defining what quantum chemists call the atom net populations. These terms simplify further. They do not contain a phase factor, and, due to the spherical symmetry of b_t, the matrix elements are non-zero only for n=m. Hence these terms reduce to

$$\langle \varrho_{\mathbf{S}} | t \rangle_w^{(1)} = \sum_n (n|t|n)_w \frac{1}{N n_{\mathbf{S}}} \sum_{\mathbf{k}\nu} n(\varepsilon_\nu(\mathbf{k})) \sum_{\mathbf{S}''}^{\{\hat{U}\mathbf{S}\}} |C_{\mathbf{S}''n;\nu}(\mathbf{k})|^2 . \tag{5.40}$$

To this expression the core electron density contribution of the site **S**

$$\langle \varrho_{\mathbf{S}} | t \rangle_w^{(c)} = \sum_c (c|t|c)_w \tag{5.41}$$

may eventually be added.

Two-centre contributions $\langle \varrho_{\mathbf{S}} | t \rangle_w^{(2)}$ arise only from the valence electron density, and they must be calculated from the full expression (5.37) excluding the entries with $\mathbf{R}+\mathbf{S}'-\mathbf{S}'' = 0$. These overlap densities $\varrho_{\mathbf{S}}^{(2)}(\mathbf{r})$ are concentrated at intermediate values of $r \approx r_{at}$ with r_{at} defined by (3.5). They are very small for $r \rightarrow 0$ as the atom-like wavefunction from the neighbouring centre $\mathbf{R}+\mathbf{S}'-\mathbf{S}'' \neq 0$ is already very small there. For this reason the expensive computation of $\langle \varrho_{\mathbf{S}} | t \rangle_w^{(2)}$ may preferably be done with a small basis $b_t^{(2)}(r) = r^4 e^{-\alpha_t r}$, $\alpha_t = 8.8/r_{WS}$, $6.6/r_{WS}$, $4.4/r_{WS}$, and the result $\varrho_{\mathbf{S}}^{(2)}(r) = \sum_t c_t b_t^{(2)}(r)$ can then eventually be reexpanded into the original basis $b_{t'}(r)$,

$$\langle \varrho_{\mathbf{S}} | t' \rangle^{(2)} = \sum_t c_t \langle t^{(2)} | t' \rangle_w ,$$

in order to have a unique site density representation.

Having obtained the matrix elements

$$\langle \varrho_{\mathbf{S}} | t \rangle_w = \langle \varrho_{\mathbf{S}} | t \rangle_w^{(c)} + \langle \varrho_{\mathbf{S}} | t \rangle_w^{(1)} + \langle \varrho_{\mathbf{S}} | t \rangle_w^{(2)} \tag{5.42}$$

the coefficients $\varrho_t^{(\mathbf{S})}$ of the site density representation (5.29) are calculated directly from (5.30). The total electronic site charge is then (with the electron charge being -1 a.u.)

$$Q_{\mathbf{S}} = -\sum_t \langle 1 | t \rangle \varrho_t^{(\mathbf{S})} , \tag{5.43}$$

where

$$\langle 1 | t \rangle = 4\pi \int_0^\infty dr\, r^2\, b_t(r) . \tag{5.44}$$

Of course, the unit cell of a crystal must be neutral, and hence

$$\sum_{\mathbf{S}} (A_{\mathbf{S}} + Q_{\mathbf{S}}) = 0 , \tag{5.45}$$

where $A_{\mathbf{S}}$ is the atomic number at the site **S**. This exact charge neutrality should be observed in any self-consistent calculation, eventually by a renormalization of the numerically obtained densities $\varrho_{\mathbf{S}}(r)$, that is of the coefficients $\varrho_t^{(\mathbf{S})}$. The total crystal electron density is then given by (5.32) as a lattice sum of overlapping site densities.

Besides the site decomposition (5.32) of the total electron density it is sometimes desirable to have a basis function representation of a spherical average of the total density within the atomic volume of the atom on site **S**:

$$\begin{aligned} \varrho(\mathbf{r}) &\approx \tilde{\varrho}_{\mathbf{S}}(|\mathbf{r}-\mathbf{S}|), \qquad |\mathbf{r}-\mathbf{S}| < r_{at}^{(\mathbf{S})}; \\ \tilde{\varrho}_{\mathbf{S}}(r) &= \sum_{t} \langle r|t\rangle \, \tilde{\varrho}_t^{(\mathbf{S})}, \qquad r < r_{at}^{(\mathbf{S})}. \end{aligned} \tag{5.46}$$

The coefficients $\tilde{\varrho}_t^{(\mathbf{S})}$ of this representation are again obtained as in (5.30),

$$\tilde{\varrho}_t^{(\mathbf{S})} = \sum \langle \tilde{\varrho}_{\mathbf{S}}|t'\rangle_{\tilde{w}} \, n_{t'}^{\tilde{w}} \, (\tilde{D}^{-1}\tilde{D}^{-1T})_{t't} \, n_t^{\tilde{w}} , \tag{5.47}$$

this time with a weighting factor

$$\tilde{w}(r) = \theta(r_{at}^{(\mathbf{S})} - r)\, w(r) \tag{5.48}$$

being non-zero only for $r < r_{at}^{(\mathbf{S})}$. With (5.32) and (5.29), the matrix elements contained in (5.47) are readily obtained as

$$\langle \tilde{\varrho}_{\mathbf{S}}|t\rangle_{\tilde{w}} = \sum_{\mathbf{RS'}} \sum_{t'} \varrho_{t'}^{(\mathbf{S'})} \langle t'_{\mathbf{R+S'}}|t_{\mathbf{S}}\rangle_{\tilde{w}} \tag{5.49}$$

with

$$\begin{aligned} \langle t'_{\mathbf{R+S'}}|t_{\mathbf{S}}\rangle_{\tilde{w}} &= \int d^3r \, \tilde{w}(r)\, b_t(r)\, b_{t'}(|\mathbf{r}-\mathbf{Q}|) = \\ &= 2\pi \int_0^{r_{at}^{(\mathbf{S})}} dr\, r^2\, w(r)\, b_t(r) \int_{-1}^{1} d\zeta \, b_{t'}(\sqrt{Q^2+r^2-2Qr\zeta}\,) \end{aligned} \tag{5.50}$$

with $\mathbf{Q} = \mathbf{R+S'-S}$. For the basis functions (5.34) the latter ζ-integral is

$$\int_{-1}^{1} d\zeta \sqrt{Q^2+r^2-2Qr\zeta}^{\,n_t-1} \exp(-\alpha_t \sqrt{Q^2+r^2-2Qr\zeta}\,) =$$

$$= \left(-\frac{\partial}{\partial\alpha_t}\right)^{n_t} \frac{1}{\alpha_t Qr} \left(e^{-\alpha_t(Q-r)} - e^{-\alpha_t(Q+r)}\right) ,$$

and it is only a matter of taste to do the r-integral of (5.50) either numerically or analytically.

6. SIMPLEX METHOD FOR k-SPACE INTEGRATION IN d DIMENSIONS

Applications of electronic structure theory for crystalline materials imply the computation of a great variety of **k**-space integrals over the Brillouin zone or its irreducible part. Examples are (5.37) and (5.40), in the case of metallic materials containing a step function $n(\varepsilon_\nu(\mathbf{k})) = \Theta(\varepsilon_F-\varepsilon_\nu(\mathbf{k}))$ in the integrand. Another class of **k**-integrals contains a δ-function $\delta(\varepsilon-\varepsilon_\nu(\mathbf{k}))$ instead of or besides the step function. The former integral might be obtained from the latter by an additional integration over ε up to ε_F. Conceptually simple but rather ineffective are histogram methods. The effectivity may be increased by combining them with interpolation schemes from a coarse mesh of **k**-points to a finer one [47, 14,97]. If the interpolation is linear, then the integral over the interpolated function may easily be calculated analytically avoiding the statistical fluctuations of histograms. Such a scheme was first introduced by Gilat and Raubenheimer [48]. A survey over the great variety of schemes of this type is given in [46]. Most popular are nowadays linear interpolation methods, using three-dimensional simplices (tetrahedrons) as domains of interpolation and first introduced independently by Lehmann and Taut [88,89] and by Jepsen and Andersen [65]. The first section of this chapter presents the results of these authors, rewritten in a way exposing the full symmetry of the algebraic terms which is most useful for programming. Integrals of the same type appear also in the two-dimensional case of surface or slab problems. Therefore the formulas are given here for the general case of d dimensions, d>1 being an arbitrary integer. This generalization, as will be seen, causes no increase of complexity of the expressions and is therefore justified. Very recently, the problem has been attacked from quite another direction [72,78]. Understanding the δ-function as the imaginary part of a pole term, the latter may be analytically continued into the physical sheet of the complex energy plane. The integrals may be computed there by standard numerics the integrand now being analytic and resulting for the integral in an analytic function of the complex variable ε. Its analytical continuation back to the real ε-axis must of course be done numerically, and a sophisticated and highly effective approach to this problem is given in full detail in [37], so that we do not repeat it here. A comparison of results of this approach with the simplex method may be found in [107]. For integrals over sufficiently isotropic **k**-dependent quantities the special direction method of Bansil [8,105] may serve as another alternative.

In section 6.2. the simplex scheme is generalized in a straightforward manner to integrals with two singular functions in the integrand. Section

6.3. illustrates on an example how the simplex method works, and a comparison with the proximity volume method for integrals of the types (5.37) or (5.40) in the case of completely occupied bands is made in the last section.

6.1. Integrals containing one singular function

First we consider the integrals

$$N(e) \equiv \int d^d k \; a(\mathbf{k}) \; \theta(e-e(\mathbf{k})), \tag{6.1}$$

$$J(e) \equiv \int d^d k \; a(\mathbf{k}) \; \delta(e-e(\mathbf{k})) = \int_{e(\mathbf{k})=e} d^{d-1} s_{\mathbf{k}} \; a(\mathbf{k}) \Big/ \left| \frac{\partial e(\mathbf{k})}{\partial \mathbf{k}} \right| = \frac{dN(e)}{de} \;, \tag{6.2}$$

$$I \equiv \oint d^d k \; a(\mathbf{k}) \; / \; e(\mathbf{k}) = \oint de \; J(e) \; / \; e \; . \tag{6.3}$$

The function N(e) is a (projected if $a(\mathbf{k}) \neq 1$) integrated density of states expression. The formulas (5.37) and (5.40) for a metal are just of this type where $n(\varepsilon_\nu(\mathbf{k})) = \theta(\varepsilon_F - \varepsilon_\nu(\mathbf{k}))$ plays the part of the singular factor. The resulting expressions (and computer codes) may of course also be used if the step function is ineffective, that is, if $\varepsilon_\nu(\mathbf{k}) < \varepsilon_F$ in the whole **k**-space as in the case of a valence band in a non-metal. J(e) is the corresponding (projected) density of states or, with $a(\mathbf{k})$ being a transition matrix element, the imaginary part of a dynamic response function. I is then up to a factor $(1/\pi)$ just the real part of the corresponding static response function, as the last expression (6.3) is just the Kramers-Kronig relation [84] for it. Of course, (6.3) may also be understood as the real part of a dynamic susceptibility I(e') at a fixed frequency e' if the functions $e(\mathbf{k})$ of (6.2) and (6.3) differ by this constant number. Then, the second Kramers-Kronig relation yields

$$J(e) = -(1/\pi^2) \oint de' \; I(e') \; / \; (e'-e) \tag{6.4}$$

which is, however, numerically less effective since I(e') is non-zero for all e' from $-\infty$ to $+\infty$ whereas J(e) of (6.2) is generally non-zero only in a finite interval of e.

In principle, if one of the functions N, J, or I is known, the other two might be obtained by differentiation or by a one-dimensional integration. In practise this may turn out not to be very useful as these functions often appear to be very singular, $J(e)$ even being a distribution. Thus, e.g. the calculation of the Fermi energy by integrating over $J(e)$ may have a very limited accuracy due to error accumulation. For these reasons, in the following, expressions from the direct **k**-space integration are given for all three of the functions. An alternative to avoid the singular functional dependencies would be to compute the functions in the physical sheet of their complex argument, the θ- and δ-functions thereby replaced by the atan and Lorentzian, respectively, and afterwards numerically to carry out an analytic continuation back to the real axis [37] as already mentioned in the introduction to this chapter.

We now consider the integrals (6.1-3) over a finite range of the d-dimensional **k**-space, the Brillouin zone or its irreducible part say, and assume this domain to be subdivided into (irregular) simplices. The shape of these simplices may be completely arbitrary, and therefore there is no problem with this subdivision. Of course, as the functions $a(\mathbf{k})$ and $e(\mathbf{k})$ are linearly interpolated in each simplex, one should avoid stretched shapes and rather keep the simplices compact with short edges (see section 6.3. for an example). All what remains is to calculate the contribution of one simplex to the integrals (6.1-3).

We denote the **k**-vectors of the corners of the simplex by $\mathbf{k}_0, \mathbf{k}_1, \ldots, \mathbf{k}_d$ and put without loss of generallity $\mathbf{k}_0 = 0$, that is, we shift the origin of the **k**-space by $\mathbf{k}_0$. Then, $\mathbf{k}_1, \ldots, \mathbf{k}_d$ are the edge vectors of the simplex, and their cartesian components are denoted by k_{ij}. The simplex volume is

$$v = |\det(k_{ij})|/d! \ , \tag{6.5}$$

and a_i, e_i, $i=0,\ldots,d$ are the function values of $a(\mathbf{k})$ and $e(\mathbf{k})$ at points $\mathbf{k}_i$. The contribution of one simplex to (6.1-3) will be denoted by $N(e;v,a_i,e_i)$ and so on, explicitly indicating all the numbers that contribution is calculated from.

To be specific, fix for a moment d=2 and consider the triangle of fig. 6.1 with corners $\mathbf{k}_0$, $\mathbf{k}_1$, $\mathbf{k}_2$. Let furthermore $e_0 < e_1 < e_2$. If $e < e_0$, there is no contribution of this simplex to $N(e)$ and $J(e)$, $N(e;v,a_i,e_i) = J(e;v,a_i,e_i) = 0$, because the integrand is zero in the whole volume of the simplex. In the case $e_0 < e < e_1$ (fig. 6.1b) the contribution to $N(e)$ is simply the volume of the occupied simplex (where $\theta(e-e(\mathbf{k})) = 1$) with edges $\mathbf{q}_i = \mathbf{k}_i(e-e_0)/(e_i-e_0)$, $i=1,\ldots,d$ multiplied by the average over the

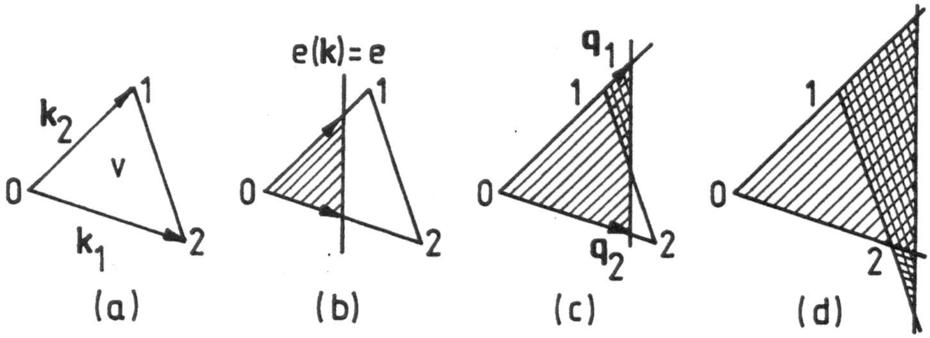

Fig. 6.1. The two-dimensional simplex with volume v (a), and the integration domain of N(e) for $e_0 < e < e_1$ (b), $e_1 < e < e_2$ (c), and $e_2 < e$ (d). See the text for explanations.

a-values corresponding to the corners of this new simplex and obtained by linear interpolation from the a_i. An elementary analysis yields

$$N(e;v,a_i,e_i) = v\, N_0(e;a_i,e_i) ,$$

$$N_0(e;a_i,e_i) = \frac{(e-e_0)^2}{\Pi_0} \left(a_0 + \frac{e-e_0}{3} \Sigma_0\right) , \qquad (6.6)$$

$$\Pi_0 = \prod_{i(\neq 0)} (e_i - e_0) , \qquad \Sigma_0 = \sum_{i(\neq 0)} \frac{a_i - a_0}{e_i - e_0} .$$

The derivative of (6.6) with respect to e is

$$J(e;v,a_i,e_i) = v\, J_0(e;a_i,e_i) ,$$

$$J_0(e;a_i,e_i) = \frac{e-e_0}{\Pi_0} (2a_0 + (e-e_0)\Sigma_0) . \qquad (6.7)$$

In the case $e_1 < e < e_2$ the domain of integration for N(e) may be composed of two simplices originating from the corners 0 and 1 and the second being subtracted from the first. The result is

$$N(e;v,a_i,e_i) = v\, [N_0(e;a_i,e_i) + N_1(e;a_i,e_i)] ,$$

$$J(e;,v,a_i,e_i) = v\, [J_0(e;a_i,e_i) + J_1(e;a_i,e_i)] , \qquad (6.8)$$

where N_1 and J_1 is obtained from the expressions N_0 and J_0, respectively, by interchanging the indices 0 and 1, and the contributions N_1 and J_1

appear to be netative because π_1 has now got a negative factor (e_0-e_1). Finally, for $e_2 < e$ the surface $e(\mathbf{k}) = e$ went through the simplex, and the contributions are simply

$$N(e;v,a_i,e_i) = \frac{v}{d+1} \sum_j a_j \ , \quad J(e;v,a_i,e_i) = 0 \ . \tag{6.9}$$

However, as seen from fig. 6.1d), the same result may likewise be obtained in the previous manner to be

$$\begin{aligned} N(e;v,a_i,e_i) &= v \sum_{j=0}^{2} N_j(e;a_i,e_i) \ , \\ J(e;v,a_i,e_i) &= v \sum_{j=0}^{2} J_j(e;a_i,e_i) \ . \end{aligned} \tag{6.10}$$

Furthermore, as seen from fig. 6.1c), (6.8) may likewise be obtained by subtracting from (6.9) the contribution N_2 and J_2, respectively, of a simplex originating from the corner 2 of the original one.

After these preliminary considerations we may immediately write down the general result for the d-dimensional case:

$$N(e;v,a_i,e_i) = v \sum_j^{(e_j<e)} N_j(e;a_i,e_i) =$$

$$= \frac{v}{d+1} \sum_j a_j - v \sum_j^{(e_j>e)} N_j(e;a_i,e_i) \ , \tag{6.11}$$

$$N_j(e;a_i,e_i) = \frac{(e-e_j)^d}{\pi_j} \left(a_j + \frac{e-e_j}{d+1} \Sigma_j\right) \ , \tag{6.12}$$

$$\pi_j = \prod_{i(\neq j)} (e_i - e_j) \ , \quad \Sigma_j = \sum_{i(\neq j)} \frac{a_i - a_j}{e_i - e_j} \ , \tag{6.13}$$

and

$$J(e;v,a_i,e_i) = v \sum_j^{(e_j<e)} J_j(e;a_i,e_i) =$$

$$= -v \sum_j^{(e_j>e)} J_j(e;a_i,e_i) \ , \tag{6.14}$$

$$J_j(e;a_i,e_i) = \frac{(e-e_j)^{d-1}}{\Pi_j} (da_j + (e-e_j)\, \Sigma_j) \ . \tag{6.15}$$

Note that the labelling of the corners of the simplex may be completely arbitrary in these expressions without any restriction on the order of the values e_i. This is especially convenient in programming.

As in the special case (6.9), $J(e;v,a_i,e_i) = 0$ for all $e > \max(e_i)$. Rearranging in this case the terms of the first line of (6.14) into a polynomial expression in powers of e we find that all coefficients of that polynomial must be zero:

$$\sum_{j=0}^{d} \frac{e_j^{n-1}}{\Pi_j} (na_j - e_j \Sigma_j) = 0 \ , \quad n = 0,\ldots,d \ . \tag{6.16}$$

These sum rules will prove to be useful in the following. In order to calculate $I(v,a_i,e_i)$ we consider first the entity

$$I_{jk} = \int_{e_k}^{e_{k+1}} de\; J_j(e;v,a_i,e_i)/e =$$

$$= \frac{1}{\Pi_j} \left[\frac{\Sigma_j}{d} (e_{k+1}^d - e_k^d) + \right.$$

$$+ \sum_{n=1}^{d-1} \binom{d}{n} \frac{(-e_j)^{n-1}}{d-n} (na_j - e_j\Sigma_j)(e_{k+1}^{d-n} - e_k^{d-n}) +$$

$$\left. + (-e_j)^{d-1} (da_j - e_j \Sigma_j) \ln \left| \frac{e_{k+1}}{e_k} \right| \right] ,$$

where (6.15) was rearranged as a polynomial in powers of e and afterwards integrated. From (6.3) and (6.14),

$$I(v,a_i,e_i) = v \sum_{k=0}^{d-1} \sum_{j=0}^{k} I_{jk} = v \sum_{j=0}^{d-1} \sum_{k=j}^{d-1} I_{jk} \; .$$

With the help of the sum rules (6.16) it is not difficult to cast the last expression into the form

$$I(v,a_i,e_i) = (-1)^d \; v \sum_{k=0}^{d} \frac{e_k^{d-1}}{\Pi_k} \{a_k + (da_k - e_k\Sigma_k) \ln|e_k|\} \; . \tag{6.17}$$

All these expressions (6.11), (6.14), and (6.17) become indefinite if two or more e_i-values coincide on one simplex. A simple and approved way of dealing with this difficulty is to split these degeneracies by an amount δe smaller than any significant e-interval and then to use the above given expressions for the non-degenerate case.

6.2. Integrals containing two singular functions

If J and I mean the imaginary and real parts of a response function of a metal, another Θ-function will appear in the integrand due to the presence of a Fermi surface. Hence, the integrals

$$J(b,e) \equiv \int d^dk \; a(\mathbf{k}) \; \Theta(b-b(\mathbf{k})) \; \delta(e-e(\mathbf{k})) \; , \tag{6.18}$$

$$I(b) \equiv \int d^dk \; a(\mathbf{k}) \; \Theta(b-b(\mathbf{k})) \; / \; e(\mathbf{k}) \tag{6.19}$$

need be considered. In close connection with (6.18) a further integral

$$K(b,e) \equiv \int d^dk \; a(\mathbf{k}) \; \delta(b-b(\mathbf{k})) \; \delta(e-e(\mathbf{k})) = \partial J(b,e)/\partial b \tag{6.20}$$

may be treated being over the intersection of two hypersurfaces in the **k**-space. It may be used, e.g., to integrate over cyclotron orbits on a Fermi-surface, yielding cyclotron masses, Dingle temperatures, and so on [95].

We consider again a single simplex and introduce a notation analogous to that used in the previous section. The trivial cases are

$$b < b_i,\ i=0,\dots,d:\quad \begin{aligned} J(b,e;v,a_i,b_i,e_i) &= 0\ ,\\ I(b;v,a_i,b_i,e_i) &= 0\ ,\\ K(b,e;v,a_i,b_i,e_i) &= 0, \end{aligned}$$

and

$$b > b_i,\ i=0,\dots,d:\quad \begin{aligned} J(b,e;v,a_i,b_i,e_i) &= J(e;v,a_i,e_i)\ ,\\ I(b;v,a_i,b_i,e_i) &= I(v,a_i,e_i)\ ,\\ K(b,e;v,a_i,b_i,e_i) &= 0\ . \end{aligned}$$

In the general case the hypersurface $b(\mathbf{k}) = b$ cuts from the considered simplex a piece, which may again be represented as a sum of signed simplices originating from the corners with $b_j < b$ and spanned by vectors

$$\mathbf{q}_{ji} = \frac{b - b_j}{b_i - b_j}(\mathbf{k}_i - \mathbf{k}_j)\ .$$

The signed volumes of these simplices are

$$\boxed{v_j = v\,\frac{(b - b_j)^d}{\prod_{i(\neq j)} (b_i - b_j)}\ ,} \tag{6.21}$$

and the interpolated function values at their corners are

$$\boxed{\begin{aligned} a_{ji(\neq j)} &= a_j + \frac{b - b_j}{b_i - b_j}(a_i - a_j)\ , \qquad a_{jj} = a_j\ ,\\ e_{ji(\neq j)} &= e_j + \frac{b - b_j}{b_i - b_j}(e_i - e_j)\ , \qquad e_{jj} = e_j\ . \end{aligned}} \tag{6.22}$$

The general results read then

$$J(b,e;v,a_i,b_i,e_i) = \sum_j^{(b_j<b)} J(e;v_j,a_{ji},e_{ji}) =$$

$$= J(e;v,a_i,e_i) - \sum_j^{(b_j>b)} J(e;v_j,a_{ji},e_{ji}) \,, \tag{6.23}$$

$$I(b;v,a_i,b_i,e_i) = \sum_j^{(b_j<b)} I(v_j,a_{ji},e_{ji}) =$$

$$= I(v,a_i,e_i) - \sum_j^{(b_j>b)} I(v_j,a_{ji},e_{ji}) \,. \tag{6.24}$$

The calculation of K(b,e) from the last expression of (6.20) is a little bit more involved. From (6.14) we see that (6.23) is composed of terms

$$v_j \; J_k(e;a_{ji},e_{ji}) \,. \tag{6.25}$$

With (6.21) and (6.22) it is not difficult to find out that the terms v_i/Π_k and Σ_k, both entering (6.25), do not depend on b. Thus, the only b-dependence of (6.25) is via the factors $(e - e_{jk})$ and via that term a_{jk} not entering Σ_k. From this fact and (6.22),

$$\frac{\partial}{\partial b} v_j \; J_j(e;a_{ji},e_{ji}) = 0 \tag{6.26}$$

immediately follows. It remains to consider the terms (6.25) with $j \neq k$. They are explicitly given by

$$v_j \; J_k(e;a_{ji},e_{ji}) = v_j \; \frac{(e - e_{jk})^{d-1}}{\tilde{\Pi}_{jk}} \left[d a_{jk} + (e - e_{jk}) \tilde{\Sigma}_{jk} \right] \,, \tag{6.27}$$

where

$$\tilde{\Pi}_{jk} = (e_j - e_{jk}) \prod_{i(\neq j,k)} (e_{ji} - e_{jk}) =$$

$$= (b - b_j)^d \, \frac{e_k - e_j}{b_k - b_j} \prod_{i(\neq j,k)} \left[\frac{e_i - e_j}{b_i - b_j} - \frac{e_k - e_j}{b_k - b_j} \right] \,, \tag{6.28}$$

and

$$\tilde{\Sigma}_{jk} = \frac{a_j - a_{jk}}{e_j - e_{jk}} + \sum_{i(\neq j,k)} \frac{a_{ji} - a_{jk}}{e_{ji} - e_{jk}} =$$

$$= \frac{a_j - a_k}{e_j - e_k} + \sum_{i(\neq j,k)} \frac{(a_i - a_j)(b_k - b_j) - (a_k - a_j)(b_i - b_j)}{(e_i - e_j)(b_k - b_j) - (e_k - e_j)(b_i - b_j)} \,. \tag{6.29}$$

To handle such expressions it is convenient to introduce combined differences

$$\boxed{\begin{aligned} D^i_{jk} &\equiv (e_i - e_j)(b_k - b_j) - (b_i - b_j)(e_k - e_j) = -D^i_{kj} \,, \\ D_{jk} &\equiv (e - e_j)(b_k - b_j) - (b - b_j)(e_k - e_j) = -D_{kj} \,. \end{aligned}} \tag{6.30}$$

Their antisymmetry with respect to the pair of subscripts is most easily seen by writing out the products. From (6.21) and (6.28) we find now

$$\frac{v_j}{\tilde{\Pi}_{jk}} = v \, \frac{(b_k - b_j)^{d-1}}{(e_k - e_j)\Pi_{jk}} \,,$$

where

$$\boxed{\Pi_{jk} = \prod_{i(\neq j,k)} D^i_{jk} \,,} \tag{6.31}$$

and

$$K_{jk}(b,e;a_i,b_i,e_i) = \frac{\partial}{\partial b} \frac{v_j}{v} J_k(e;a_{ji},e_{ji}) =$$

$$= -d \, \frac{(b_k - b_j)^{d-2}(e - e_{jk})^{d-2}}{\Pi_{jk}} \Bigg[(d-1)a_{jk} +$$

$$+ (e - e_{jk})\left(\tilde{\Sigma}_{jk} - \frac{a_k - a_j}{e_k - e_j}\right)\Bigg] \,. \tag{6.32}$$

Inserting here (6.29) and using

$$(d-1)a_{jk} = \sum_{i(\neq j,k)} a_{jk} ,$$

the final expression

$$K_{jk}(b,e;a_i,b_i,e_i) = -d \frac{(D_{jk})^{d-2}}{\Pi_{jk}} \Sigma_{jk} ,$$

$$\Sigma_{jk} = \sum_{i(\neq j,k)} \frac{1}{D^i_{jk}} \left[a_i D_{jk} + a_k D_{ij} + a_j D_{ki}\right] \tag{6.33}$$

is obtained displaying via (6.30) the symmetry properties

$$K_{jk}(b,e;a_i,b_i,e_i) = -K_{kj}(b,e;a_i,b_i,e_i) =$$

$$= -K_{jk}(e,b;a_i,e_i,b_i) . \tag{6.34}$$

Combining now (6.20), (6.23), (6.14), and (6.32) yields

$$K(b,e;v,a_i,b_i,e_i) = v \sum_{j\neq k}^{(b_j<b)(e_k<e)} K_{jk}(b,e;a_i,b_i,e_i)$$

where, however, in the sum terms may cancel due to (6.34) so that the result reduces further to

$$K(b,e;a_i,b_i,e_i) = v \sum_{j\neq k}^{\substack{(b_j<b)(e_k<e)\\ \text{and } (b<b_k) \text{ or } (e<e_j)}} K_{jk}(b,e;a_i,b_i,e_i) . \tag{6.35}$$

By the definition (6.20) K must be symmetric with respect to interchanging all b-values with all e-values. This symmetry is readily seen from (6.35) if one simultaneously interchanges the summation indices and uses (6.34).

Again, (6.23) and (6.24) turn indefinite if at least two of the b_i-values or two of the e_i-values are equal to each other on one simplex. Analogously, (6.35) turns indefinite in the case of a degeneracy of at least three values b_i or e_i. All these complications may again be circumvented by lifting the degeneracies by an unsignificantly small amount.

6.3. An example

As an example we present results of a density of states calculation for the free-electron bandstructure of the empty f.c.c. lattice via (6.13-15) with

$$a(\mathbf{k}) = 1 , \quad e(\mathbf{k}) = \varepsilon_\nu(\mathbf{k}) = (\mathbf{k} - \mathbf{G}_\nu)^2 , \tag{6.36}$$

and d=3. The expression (6.36) for $\varepsilon_\nu(\mathbf{k})$ was used in the irreducible part of the Brillouin zone shown in fig. 6.2 over which the integral (6.2) was taken assuming $\varepsilon_\nu(\mathbf{k})$ to have the full point symmetry of the reciprocal lattice. For $\mathbf{G}_\nu$ a few of the smallest reciprocal lattice vectors were used,

$$\begin{aligned} &\mathbf{G}_0 = 0, \quad \mathbf{G}_1 = (1,1,1), \quad \mathbf{G}_2 = (2,0,0), \quad \mathbf{G}_3 = (1,1,-1), \\ &\mathbf{G}_4 = (0,2,0), \quad \mathbf{G}_5 = (1,-1,1), \quad \mathbf{G}_6 = (1,-1,-1), \end{aligned} \tag{6.37}$$

so that everywhere $\varepsilon_0 < \varepsilon_1 < \varepsilon_2 < \ldots$, and the density of states result should accumulate to

$$D(e) \sim \sqrt{e} \tag{6.38}$$

when summed over ν.

The used subdivision of the irreducible part of the Brillouin zone into tetrahedrons is given in fig. 6.2. The set of all corners of all tetrahedrons forms a regular simple cubic mesh. In order that this mesh fits into the domain of integration, the number of intervals on the line Γ - X must be a multiple of four. The accuracy of interpolation within a simplex depends on its diameter rather than on its volume. As is well known, the diameter of a simplex is equal to the length of its longest edge. Hence, subdividing the space into simplices, it is their edges rather than their volumes which must be kept as small as possible. For this reason, a mesh cube of fig. 6.2 was subdivided into five tetrahedrons of equal diameter rather than into six tetrahedrons of equal volume. The stretched tetrahedron number 6 of fig. 6.2 is symmetry equivalent to the two halves of the tetrahedrons number 22 and 23 lying nominally outside of the first Brillouin zone. Therefore the former may be omitted by including the latter[1].

---

[1] This detail was pointed out to the author by A. C. Switendick. The autor benefitted from private discussions with A. C. Switendick particularly regarding the content of this section and the next.

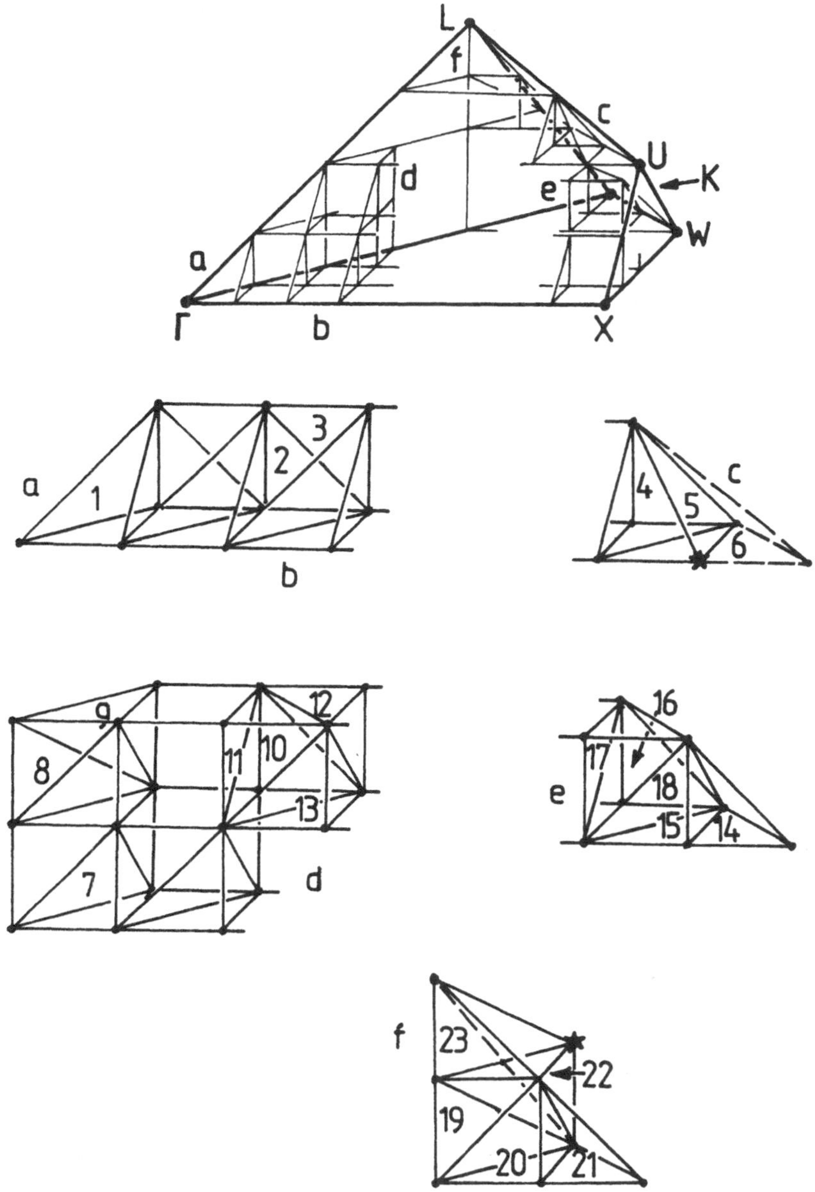

Fig. 6.2. The irreducible part of the Brillouin zone of the f.c.c. lattice and its subdivision into tetrahedrons. There are 23 types of tetrahedrons different by shape or orientation. Tetrahedron 6 is to be omitted in favour of the halves of the tetrahedrons 22 and 23 lying nominally outside of the Brillouin zone but being symmetry equivalent to tetrahedron 6.

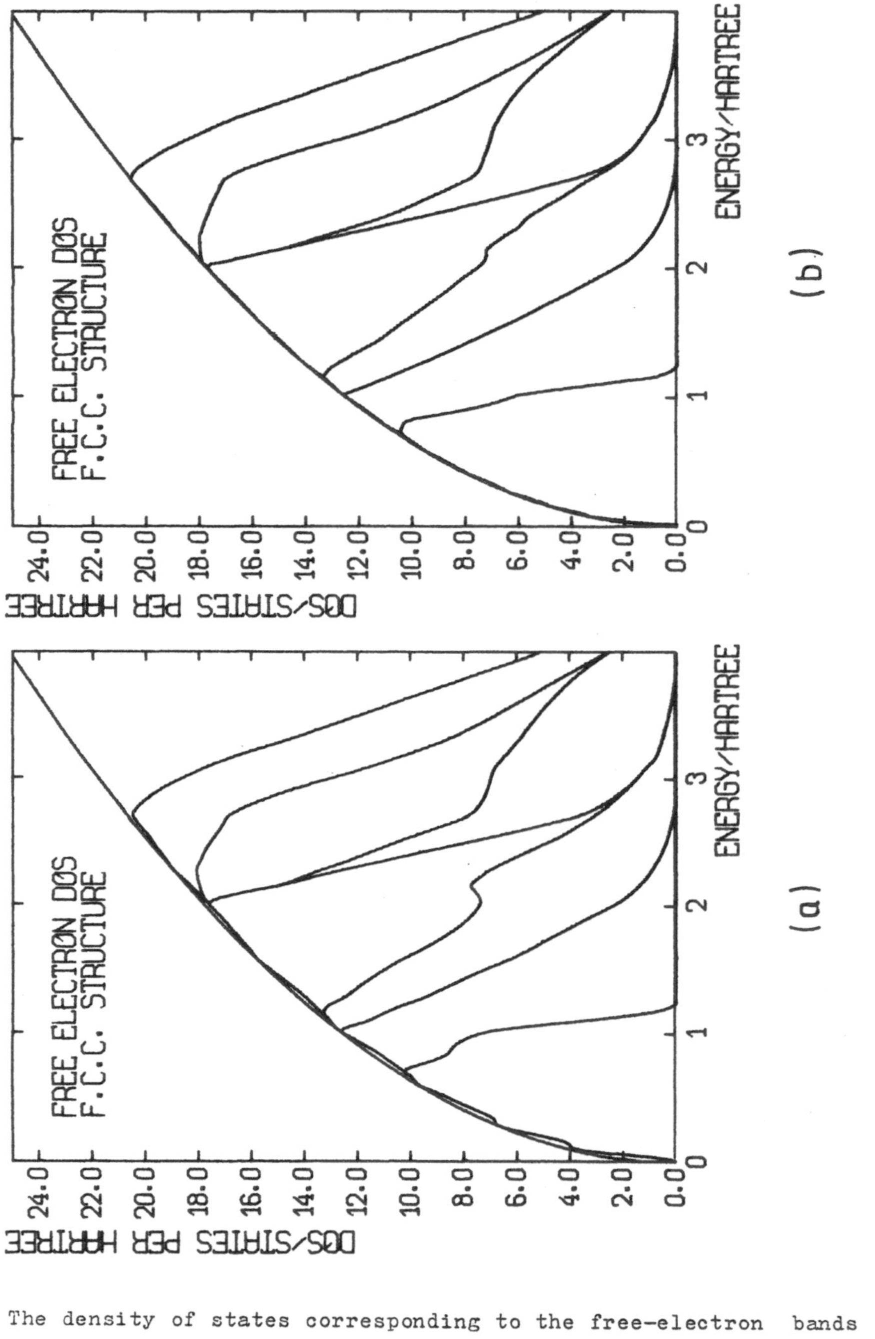

Fig. 6.3. The density of states corresponding to the free-electron bands (6.36-37) for the f.c.c. structure, obtained by simplex integration over the irreducible part of the Brillouin zone, subdivided as shown in fig. 6.2 with four (a) and eight (b) intervals on the line $\Gamma - X$, respectively. The exact free-electron parabola (6.38) is shown for comparison.

The results for the density of states are given on fig. 6.3. It is our experience with density of states-like problems and with problems of the type (5.37) or (5.40) that eight intervals on the line $\Gamma - X$ are sufficient for any purpose, in many cases even four intervals do.

6.4. Comparison with the proximity volume method

The simplex integration method was recently criticized by Kleinman [71]. He considered the bandstructure energy

$$E_{bs} = \frac{V_{U.C.}}{(2\pi)^3} \int d^3k \; e(\mathbf{k}) \tag{6.39}$$

for a hypothetical band

$$e(\mathbf{k}) = (\cos x_1 \; \cos x_2 \; \cos x_3)^2 \; , \quad x_i = \frac{\pi k_i}{a} \; . \tag{6.40}$$

in the Brillouin zone of the b.c.c. lattice. This function has in fact the periodicity of a simple cubic Brillouin zone of half of the linear size of the zone corresponding to the b.c.c. lattice, and E_{bs} is therefore easily obtained by analytically integrating over a cube giving

$$E_{bs}^{exact} = 0{,}125.$$

The simplex method (which reduces in this simple case to (6.9)) with a mesh of 14 **k**-points within the irreducible wedge of the b.c.c. Brillouin zone (four intervals on the line $\Gamma - H$) yields $E_{bs}^{s} = 0.143$, whereas a proximity volume integration, taking the proximity volume of the **k**-point as a weight which corresponds to the simple equal weight summation over the full period of $e(\mathbf{k})$, reproduces the exact result: $E_{bs}^{pv} = 0.125$.

Kleinman observes that, if the result of applying the simplex method to this case is rewritten as

$$E_{bs}^{s} = \sum_i w_i \; e(\mathbf{k}_i) \; , \tag{6.41}$$

then, on the line $\Gamma - H$, symmetry equivalent points $\mathbf{k}_i$ may get different weights w_i, and he qualifies this as a defect of the method not present e.g. in the proximity volume method. In fact, however, there are no general restrictions on the weights w_i of an integration formula like

(6.41). Of course, for any set of w_i there is a class of functions $e(\mathbf{k})$ for which (6.41) reproduces the exact result. The example (6.40) by Kleinman just falls in that class for the proximity volume formula. If he would have chosen, e.g.,

$$e(\mathbf{k}) = \left(\frac{\pi}{a}\,\mathbf{k}\right)^2, \qquad (6.42)$$

he would have obtaied

$$E_{bs}^{exact} = \frac{9}{20} = 0.45, \quad E_{bs}^{s} = 0.406, \quad E_{bs}^{pv} = 0.384$$

for the same set of 14 **k**-points. As the proximity volume method is based on zeroth order interpolation and the simplex method on first order interpolation, the error of the former will <u>on average</u> exceed that of the latter.

This last statement is, however, not always useful if the application aims at a special class of functions as in the case of the bandstructure energy (6.39) of a single isolated band $e(\mathbf{k})$ or when $e(\mathbf{k})$ is replaced by the sum $\sum_\nu \varepsilon_\nu(\mathbf{k})$ of an isolated band complex, the valence bands of a semiconductor say. Here, $e(\mathbf{k})$ is periodic and analytic.

Consider for a moment the one-dimensional case and the integral

$$I_n = \int_0^1 dx \cos(2\pi nx) = \delta_{n0} . \qquad (6.43)$$

If this integral is computed with the equal step equal weight (proximity volume) formula from function values at points $x_i = i/m$, $1 < i < m$, the result is

$$I_n^{pv,m} = \frac{1}{m} \sum_{j=1}^{m} \cos \frac{2\pi nj}{m} . \qquad (6.44)$$

Let now $n = r \bmod m$, that is $n = km + r$, $r < m$. For $r=0$ we have obviously $I_n^{pv,m} = 1$ because all cosines of (6.44) are equal to unity in this case. For $0 < r < m$ we find from the real part of

$$\sum_{j=1}^{m} \left(e^{i2\pi r/m}\right)^j = e^{i2\pi r/m} \frac{1 - e^{i2\pi r}}{1 - e^{i2\pi r/m}} = 0$$

that $I_n^{pv,m} = 0$, hence,

$$I_n^{pv,m} = \delta_{r0} , \quad n = r \bmod m . \tag{6.45}$$

Comparing this to (6.43) we see that the equal step proximity volume result has got an error only in the case $n = km \neq 0$.

Comming back to our bandstructure energy in the tight-binding case where

$$\sum_{\nu} \varepsilon_{\nu}(\mathbf{k}) = \mathrm{tr}\, \mathbf{H}(\mathbf{k}) = \sum_{\mathbf{R}}^{|\mathbf{R}|<R} E_{\mathbf{R}} \cos(\mathbf{kR}) , \tag{6.46}$$

we see that using the proximity volume method with a regular mesh of points $\mathbf{k}_i$ we obtain the exact result for the bandstructure energy, if only in each coordinate direction the number of points exceeds the largest number of periods of an entry $\cos(\mathbf{kR})$, contained in the Brillouin zone. In the example (6.40) this latter number is two in the b.c.c. Brillouin zone. Therefore three points on the line $\Gamma - H$, that is the coarser mesh of in total five points in the irreducible wedge would have sufficed in this case to reproduce the exact result by the proximity volume method.

All these considerations are, however, of no use if integrals of the types (6.1-3) or (6.18-20) are to be treated with singular functions in the integrand.

7. POTENTIAL CONSTRUCTION AND ITERATION

In this closing chapter the construction of the effective crystal potential (in the sense of (1.26) or (1.78); for the sake of definiteness we concentrate on (1.26)) in terms of a lattice sum (2.63) of site potentials from the calculated electron density (cf. chapter 5.) and the organization of the self-consistency iterations for the Kohn-Sham equations (1.26) are described. There are several requirements to be met by the potential representation in order to be effective for an LCAO treatment as described in the present work. These requirements are however much weaker than those imposed by the commonly used muffin-tin form of the crystal potential; they are discussed in the first section. The site representation of the Coulomb part v_H of the potential is, due to the linearity of Poisson's equation, readily obtained from the site representation of the charge density given in chapter 5. The only problem here is the treatment of ionicity. All the explicit formulas needed for the Hartree potential v_H to be obtained in a basis function expansion are derived in the second section of this chapter. The third section deals with the same representation for the exchange and correlation potential v_{XC} of (1.26) depending on the electron density in a non-linear way.

Exactly the same treatment as described here of course might be used for a local approximation v_{Σ} of the self-energy in (1.78), e.g. that of type (1.83). An eventual energy dependence of v_{Σ} could preferably be linearized around ε_F the corresponding LCAO matrix elements of the linear term being then added to the B-matrix (2.53). Such a treatment would cause very little increase in computing time against that for an energy-independent potential

The chapter closes with the description of an effective approach to the self-consistency iterations. This procedure is distinctly superior to the commonly used simple potential mixing procedure, and it can be used for the iterative solution of any non-linear vector problem of high dimension, where the function calculation is very expensive.

7.1. Principles of potential construction

The self-consistent crystal potential $v(\mathbf{r}) = v_H(\mathbf{r}) + v_{XC}(\mathbf{r})$ entering (1.26) is connected with the total ionic and electronic charge density of the crystal via (1.18) (in which formula v means the Coulomb potential of the nuclei, whereas in this chapter v is the total effective crystal potential) and (1.29), (1.37). A great part of existing bandstructure calculation methods requires a potential representation

$$v(\mathbf{r}) = v_{mt}(\mathbf{r}) + v'(\mathbf{r}) \tag{7.1}$$

where

$$v_{mt}(\mathbf{r}) = \begin{cases} v_{mt}^{S}(|\mathbf{r} - \mathbf{R} - \mathbf{S}|) & \text{for} \quad |\mathbf{r} - \mathbf{R} - \mathbf{S}| < r_{mt}^{(S)} \\ \text{const.} & \quad\quad \text{elsewhere,} \end{cases} \tag{7.2}$$

and v' is weak enough to be treated in first order perturbation theory. This representation has proved to give excellent results in the case of close-packed metallic structures even when neglecting v' [96], whereas v' must be included in the case of open structures (at least if no interstitial sites are introduced; cf. the end of section 3.2.).

LCAO methods usually impose weaker restrictions onto the potential representation. In our treatment we need a representation (2.63), preferably with spherical site potentials, that is

$$v(\mathbf{r}) = \sum_{\mathbf{RS}} v_{S}(|\mathbf{r} - \mathbf{R} - \mathbf{S}|) + \text{const.} \tag{7.3}$$

where, however, contrary to (7.2) the site potentials may heavily overlap. Of course, the numerical effort increases with increasing range of the site potentials. Thus the strategy must be to find a representation (7.3) as close to the true crystal potential v(**r**) as possible but with spherical potential wells v_S as localized as possible. Note also that the potential constant in (7.3) has a physical meaning only in connection with surface properties (work function and so on) and may be put to zero in the course of the band calculation for the bulk.

The ingredients of the potential calculation are, as already mentioned, the nuclear charges and positions and the total electronic charge density. As the connection (1.18) between the charge density and the Hartree potential $v_H(\mathbf{r})$ is linear, the representation (5.32) of the electron density readily yields the Hartree potential in the form (7.3). If there are, however, several atom sorts per unit cell, then the total electronic charge Q_S per atom given by (5.43) does no longer neutralize the nuclear charge A_S. This ionicity would lead to an infinite range of the site potential v_S of (7.3) and needs therefore a special treatment.

The connection (1.29), (1.37) between the electronic number density $\varrho(\mathbf{r})$ and the exchange and correlation potential v_{XC} is non-linear, and hence the representation (5.32) is not cooperative in this case. In order to maintain the spherical symmetry of $v_S(r)$ in (7.3) we use an atomic sphere approximation (5.46) for the total electron density resulting in a corresponding atomic sphere approximation for v_{XC}. The exchange and correlation potential being roughly proportional to the (1/3) power of the density varies much less than the density itself in the interstitial region, and an atomic sphere approximation should be justified at least if one works in the case of open structures with additional interstitial sites S having their own contributions ϱ_S.

7.2. The Hartree part of the potential

In order to separate the ionicity problem mentioned in the last section the total site charge density of the atom S in the crystal is split according to

$$\eta_S(r) = [Q_S\delta(r) - \varrho_S(r)] + (A_S - Q_S)\,\delta(r)\ , \tag{7.4}$$

where Q_S is given by (5.43), $\varrho_S(r)$ is the electron site density entering (5.32), and A_S is the nuclear charge of the atom S. The square brackets of (7.4) define a neutral entity. If the mechanical potential acting on an electron (that is, in our units, minus the electrostatic potential) is written down as in (2.20),

$$v_{H,S}(r) = -\frac{1}{r}\, z_H^{(S)}(r)\ , \tag{7.5}$$

then the contribution of the terms in square brackets of (7.4) to $z_H^{(S)}$ (denoted by $z_n^{(S)}$, n reminding "neutral") is

$$z_n^{(S)}(r) = \int_r^\infty dr' \int_{r'}^\infty dr''\ 4\pi r''\ \varrho_S(r'')\ . \tag{7.6}$$

This corresponds to the solution of Poisson's equation for $v_{H,S}(r)$ with boundary conditions

$$z_n^{(S)}(\infty) = (dz_n^{(S)}/dr)(\infty) = 0\ . \tag{7.7}$$

The potential will finally again be expanded into basis functions according to (2.20-25), where $z^{(S)}(r)$ enters only in the matrix elements $\langle z^{(S)}|t\rangle_w$ with w = 1 as found at the end of section 2.1. to be the best choice. Using on the other hand the expansion (5.29) for $\varrho_S(r)$, we are left with

$$\langle z_n^{(S)}|t\rangle_w = 4\pi \sum_{t'} \varrho_{t'}^{(S)} \int_0^\infty dr\; r^{n_t-1} e^{-\alpha_t r} \int_r^\infty dr' \int_{r'}^\infty dr''\; r''^{\,n_{t'}} e^{-\alpha_{t'} r''}$$

$$= 4\pi \sum_{t'} \varrho_{t'}^{(S)} \frac{n_{t'}!}{\alpha_{t'}^{n_{t'}+2}} \sum_{m=0}^{n_{t'}} (n_{t'}+1-m) \frac{(n_t-1+m)!}{(\alpha_{t'}+\alpha_t)^{n_t+m}} \frac{\alpha_{t'}^m}{m!}$$

(7.8)

giving the final answer for the "neutral part" of the Coulomb potential.

The last term of (7.4) vanishes in the case of a homoatomic crystal. Otherwise it demands the application of an Ewald technique [39]: First this term is again split according to ($Z_S = (A_S - Q_S)$)

$$Z_S\, \delta(r) = Z_S \left[\delta(r) - \frac{p^3}{\pi^{3/2}} e^{-p^2 r^2}\right] + Z_S \frac{p^3}{\pi^{3/2}} e^{-p^2 r^2} \qquad (7.9)$$

with the Ewald parameter p yet to be determined. Formula (7.9) adds and subtracts a normalized Gaussian charge density to the point charge so that the entity in square brackets again is a neutral one. It is treated in the same way as above resulting in a contribution $z_i^{(S)}$ (i reminding "ionic") to $z_H^{(S)}$:

$$z_i^{(S)}(r) = Z_S\, \mathrm{erfc}(pr) \qquad (7.10)$$

and

$$\langle z_i^{(S)}|t\rangle_w = \begin{cases} \dfrac{Z_S}{t}\left[1 - e^{a_t^2}\, \mathrm{erfc}\, a_t\right] & \text{for } n_t = 1 \\[2ex] \dfrac{Z_S}{t^2}\left[1 + \dfrac{2}{\sqrt{\pi}} a_t - (1+2a_t^2)\, e^{a_t^2}\, \mathrm{erfc}\, a_t\right] & \text{for } n_t = 2 \\[2ex] \dfrac{2Z_S}{t^3}\left[1 + \dfrac{2}{\sqrt{\pi}} a_t^3 - (1+a_t^2+2a_t^4)\, e^{a_t^2}\, \mathrm{erfc}\, a_t\right], & n_t = 3 \end{cases}$$

$$a_t \equiv \alpha_t / 2p.$$

(7.11)

It remains the last term of (7.9). This term is Fourier transformed into

$$\tilde{\varrho}_S(q) = Z_S \frac{p^3}{\pi^{3/2}} \int d^3r \; e^{-iqr} \; e^{-p^2r^2} = Z_S \; e^{-q^2/4p^2} \; . \tag{7.12}$$

The Fourier transform of the corresponding Coulomb potential is

$$\tilde{v}_S(q) = \frac{4\pi\tilde{\varrho}_S(q)}{q^2} = \frac{4\pi \; Z_S}{q^2} \; e^{-q^2/4p^2} \; . \tag{7.13}$$

The total potential contribution of this origin to the crystal potential is

$$v'(r) = \sum_{RS} \frac{1}{(2\pi)^3} \int d^3q \; e^{iq(r-R-S)} \; \tilde{v}_S(q) \; . \tag{7.14}$$

With (cf. e.g. [122])

$$\sum_R e^{-iqR} = \frac{(2\pi)^3}{V_{U.C.}} \sum_G \delta(q - G) \tag{7.15}$$

and (7.13) the result (7.14) may be rewritten as

$$v'(r) = \sum_G e^{iGr} \; \tilde{v}_G \; ,$$

$$\tilde{v}_G = \sum_S \frac{4\pi \; Z_S}{V_{U.C.}} \; e^{-iGS} \; e^{-G^2/4p^2} \; . \tag{7.16}$$

The zero momentum Fourier component vanishes due to (5.45),

$$\tilde{v}_0 = \frac{4\pi}{V_{U.C.}} \sum_S Z_S = 0 \; . \tag{7.17}$$

(Strictly speaking, for an infinite crystal (7.17) is a definition due to the divergency of the sum (7.14).) The point is now the following: If we put

$$\boxed{\frac{1}{p} = \frac{3}{4} R_{min} \, ,} \tag{7.18}$$

where R_{min} is chosen so that for $\mathbf{G} \neq 0$ the estimate $|\mathbf{G}| R_{min} \geq 2\pi$ holds, i.e. in most cases, R_{min} be approximately the modulus of the smallest lattice vector, then

$$e^{-G^2/4p^2} \leq e^{-9\pi^2/16} \approx 4 \cdot 10^{-3} \, . \tag{7.19}$$

Because for all crystals with more than one atom per unit cell $4\pi / V_{U.C.} < 0.1$ and $Z_S \sim 1$, we find that for the choice (7.18) a single item of (7.16) will not exceed 1 mRyd and $v'(\mathbf{r})$ will in total amount only a few mRyd and hence may completely be neglected. On the other hand, with the same choice (7.18) we find from (7.10)

$$z_i^{(S)}(R_{min}) = Z_S \, \mathrm{erfc}(4/3) = 0.059 \, Z_S \, , \tag{7.20}$$

that is, the total Coulomb potential (7.5) with

$$z_H^{(S)} = z_n^{(S)} + z_i^{(S)} \tag{7.21}$$

is rather well localized.

7.3. The exchange and correlation potential

As already mentioned, the exchange and correlation potential is calculated from the atomic sphere approximation (5.46) of the total electron density. Here, the first question to be raised is how to determine the atom radii $r_{at}^{(S)}$. In a homoatomic crystal there is only one such number which is simply determined by the volume of the unit cell:

$$\frac{4\pi}{3} \sum_S [r_{at}^{(S)}]^3 = V_{U.C.} \, . \tag{7.22}$$

In a heteroatomic crystal the ratios of the atom radii remain indetermined by this condition. They are suitably fixed by the conditions

$$\tilde{\varrho}_{S_1}(r_{at}^{(S_1)}) = \tilde{\varrho}_{S_2}(r_{at}^{(S_2)}) = \ldots = \tilde{\varrho}_{S_n}(r_{at}^{(S_n)}) \equiv \bar{\varrho} \tag{7.23}$$

with $\tilde{\varrho}_S(r)$ from (5.46). With (7.22-23) we have just n non-linear equations for the n unknown numbers $r_{at}^{(S)}$, which may easily be solved iteratively because the functions $\tilde{\varrho}_S(r)$ are very smooth. Computing $\tilde{\varrho}_S$ via (5.47), it is useful in calculating R and $<\tilde{\varrho}_S|t>_w$ to choose a somewhat larger value $r_{at}^{(S)}$ than the expected solution of (7.22-23) in order to be in the position to use the once calculated function $\tilde{\varrho}_S(r)$ during the iterative process.

The site contribution to the exchange and correlation potential of the crystal is now readily obtained via (1.29) and (1.37) as

$$\boxed{v_{XC,S}(r) = \begin{cases} v_{XC}(\tilde{\varrho}_S(r)) - v_{XC}(\bar{\varrho}) \\ 0 \end{cases} \quad \text{for} \quad r \ \begin{matrix} < \\ > \end{matrix} \ r_{at}^{(S)} .} \tag{7.24}$$

The subtraction of the constant $v_{XC}(\bar{\varrho})$ (which thus enters the unimportant constant term of (7.3)) transforms $v_{XC,S}$ into a <u>continuous</u> site potential of finite range which even has (due to the flatness of $\tilde{\varrho}_S(r)$ at $r = r_{at}^{(S)}$) an almost continuous derivative and hence may be well expanded into the representation (2.20). Furthermore, the subtraction in (7.24) makes it possible to use this form with overlapping atomic spheres. The summation of (small) site terms (7.24) in the overlap region yields a correction in the right direction since the true crystal density $\varrho(\mathbf{r})$ is generally to be expected somewhat larger than $\tilde{\varrho}_S$ in the overlap (bond) region. Anyhow, the crystal potential representation (7.3) with

$$v_S(r) = v_{H,S}(r) + v_{XC,S}(r) \tag{7.25}$$

given by (7.5), (7.21), and (7.24) is at least more accurate than (7.2).

Due to the complicated functional dependence of $v_{XC}(\varrho)$ on ϱ the matrix element

$$<z_{XC}^{(S)}|t>_w = <-rv_{XC,S}|t>_w \tag{7.26}$$

must be computed numerically. This calculation contains one subtle point: Due to the strong non-orthogonality of the used basis $\{b_t\}$ of the type (2.10) (resulting in strong oscillations of the corresponding orthogonalized basis functions) the computation of the matrix elements (7.26) with the various basis functions must be <u>consistent</u> in order to avoid spurious oscillations in the basis representation of the potential:

$$\boxed{\begin{aligned} \langle z_{XC}^{(S)}|t\rangle_w &\equiv -\int_0^\infty dr\ r\ v_{XC,S}(r)\ b_t(r) = \\ &= \sum_{i=1}^{m} \int_{r_{i-1}}^{r_i} dr\ (z_{i-1} + r(z_i - z_{i-1}))\ b_t(r)\ , \\ r_0 = 0,\quad r_m &= r_{at}^{(S)},\quad z_i = -r_i\ v_{XC,S}(r_i)\ , \end{aligned}} \tag{7.27}$$

where the basis function integrals with the interpolated function $z_{at}^{(S)}(r)$ are now calculated analytically. Hence, the interpolated function is exactly expanded. (The same intervals (r_{i-1}, r_i) are to be used for all basis functions.)

By these considerations the potential construction is completed. Now, putting

$$\langle z^{(S)}|t\rangle_w = \langle z_n^{(S)}|t\rangle_w + \langle z_i^{(S)}|t\rangle_w + \langle z_{XC}^{(S)}|t\rangle_w \tag{7.28}$$

into (2.24-25) (with $A = A_S$ and $\beta_t = b_t(0)$) the basis representation (2.20) of the total site potential (7.25) is obtained. An approved basis at site S is

$$\begin{aligned} &b_t(r) \equiv b_{\nu\alpha}(r) = r^{\nu-1}\ e^{-\alpha r}\ , \\ &\nu = 1,2,3;\quad \alpha = A_S,\ A_S/q,\ \dots\ ,\ 4.4/r_{at}^{(S)};\quad q \approx 3, \end{aligned} \tag{7.29}$$

being the same as for the s-wavefunction expansion.

7.4. Iteration of a high-dimensional non-linear vector equation[1]

The most naive iteration procedure of a self-consistent bandstructure calculation would be the following: Take any reasonable crystal potential, in our scheme given, e.g., by the numbers v_t of (2.20) for the various lattice sites S, or, preferably as will soon be seen (cf. (7.31)), the numbers

$$\overline{v}_t \equiv \sum_{t'} R_{tt'}\ (v_{t'}/n_{t'}) \tag{7.30}$$

[1] The procedure described in this section was developed by A. Möbius and H. Eschrig (unpublished). The main idea is due to A. Möbius, whereas the present author speeded up the procedure by nearly a factor of two by adding the last term to (7.37). (Cf. also [50].)

with $\mathbf{R}$ from (2.18) and n_t from (2.16), being the expansion coefficients into an orthonormalized basis at site $\mathbf{S}$. We join all the coefficients $\bar{v}_t$ for all $\mathbf{S}$ into a high-dimensional vector $\mathbf{X}_1$ (typically of dimension 10 to 100). With this potential perform a bandstructure calculation, calculate the electron density, and get as the result a new potential $\mathbf{X}_2 = \mathbf{G}(\mathbf{X}_1)$. With the choice (7.30) for the vector components of $\mathbf{X}$ the difference vector immediately gives

$$\mathbf{Y}_1^2 = (\mathbf{G}(\mathbf{X}_1) - \mathbf{X}_1)^2 = \sum_{\mathbf{S}} \int_0^\infty dr\, r^2\, (v_{new}(r) - v_{old}(r))^2. \qquad (7.31)$$

Repeat the procedure up to convergence $\mathbf{Y}^2 \longrightarrow 0$.

Unfortunately this procedure will almost never converge, and what most people do is a simple mixing, that is, $\mathbf{X}_2$ is replaced by

$$\mathbf{X}_2' = \mathbf{X}_1 + p\mathbf{Y}_1 \qquad (7.32)$$

with $p < 1$ chosen by experience. The convergence enforced in this way may be very slow [104, 27], and more advanced procedures are desired [28].

Independently of scf-electronic structure calculations this is a standard numerical problem: solve by iterations the high-dimensional non-linear vector equation

$$\mathbf{X} = \mathbf{G}(\mathbf{X}) \ , \qquad (7.33)$$

where the computation of $\mathbf{G}(\mathbf{X})$ for a given $\mathbf{X}$ is extremly costly. The problem (7.33) is equivalent to the extremum problem

$$\mathbf{Y}^2 = F(\mathbf{X}) \implies \min. \qquad (7.34)$$

Standard numerics uses gradient methods in this case (eventually replacing the gradient by finite differences), however, to compute the gradient to F by finite differences one needs (n+1) calculations of $\mathbf{G}(\mathbf{X})$. We are looking for a procedure which with a less number of calculations of $\mathbf{G}(\mathbf{X})$ hopefully has already converged.

Suppose that we have already performed (m+1) calculations of

$$\mathbf{Y}_i = \mathbf{G}(\mathbf{X}_i) - \mathbf{X}_i \ , \quad i = 0, \ldots, m.$$

By linear interpolation in the m-dimensional subspace spanned by the $\mathbf{X}_i$ we obtain

$$\mathbf{Y}(\mathbf{g}) = \sum_{i=0}^{m} g_i \, \mathbf{Y}_i \,, \qquad \mathbf{X}(\mathbf{g}) = \sum_{i=0}^{m} g_i \, \mathbf{X}_i \,, \tag{7.35}$$

and

$$F(\mathbf{g}) = (\mathbf{Y}(\mathbf{g}))^2 = \sum g_i g_j \, (\mathbf{Y}_i, \mathbf{Y}_j) \,.$$

We may find the minimum of $F(\mathbf{g})$ in this subspace

$$F(\mathbf{g}) \implies \min, \qquad \frac{\partial F}{\partial g_i} = 2 \sum_{j=0}^{m} g_j \, (\mathbf{Y}_i, \mathbf{Y}_j) = 0$$

by solving the linear equation system

$$\boxed{\sum_{j=0}^{m} g_j \, (\mathbf{Y}_i, \mathbf{Y}_j) = 0} \tag{7.36}$$

for $\mathbf{g}$. Putting this solution into (7.35) we find the vectors $\mathbf{X}$ <u>and</u> $\mathbf{Y}$ at the minimum in the given subspace, and in analogy to (7.32) we may put

$$\mathbf{X}_{m+1} = \sum_{i=0}^{m} g_i \, \mathbf{X}_i + p_m \left(\sum_{i=0}^{m} g_i \, \mathbf{Y}_i \right) , \tag{7.37}$$

$$\mathbf{Y}_{m+1} = \mathbf{G}(\mathbf{X}_{m+1}) - \mathbf{X}_{m+1} \,. \tag{7.38}$$

In general (this is one of the necessary conditions for convergence of the method) $\mathbf{Y}(\mathbf{g})$ will point out of the m-dimensional subspace spanned by the $\mathbf{X}_i$, $i = 0, \dots, m$, so that the new set of $\mathbf{X}_i$ completed by $\mathbf{X}_{m+1}$ span already an (m+1)-dimensional subspace. Furthermore, whereas in (7.32) experience was needed to choose p, in (7.37) the experience of the last iteration step may be used to determine p_m:

$$\boxed{p_m = \left(\sum_{i=0}^{m} g_i \mathbf{X}_i - \mathbf{X}_m, \, \mathbf{Y}_m \right) / \mathbf{Y}_m^2 \,.} \tag{7.39}$$

The geometrical sense of this choice is easily understood.

Some points are yet to be considered before running the iterations. First of all, in the course of iterations, Y^2 will decrease by several orders of magnitude. Thus, approaching the solution of (7.33/34), the far away lying points X_i become less and less useful in the linear interpolation (7.35) of $Y(X)$. This may easily be controlled by comparing the magnitudes of

$$\left| \sum_{i=0}^{m} g_i \, Y_i \right| \quad \text{and} \quad |Y_{m+1}| \; .$$

If the latter quantity exceeds the former considerably, the worst points X_i (with largest $|Y_i|$) must be dropped. Another danger is an approaching degeneracy of the equation system (7.36). This may be controlled by checking the ratio of the smallest diagonal element in the Cholesky decomposition of the matrix (Y_i, Y_j) with the smallest $|Y_i|$. If this ratio underflows a certain value, 0.1 say, then again some of the worst X_i must be dropped. (This test works, if the Y_i are ordered with descenting $|Y_i|$ which is convenient anyhow.) Eventually a new direction for the minimum search may be opened by randomly changing the direction of $\sum g_i Y_i$ in (7.37). This should, however, not be done without necessity. Finally, the maximum dimension m of the searching space may be limited due to a limited computer storage. Thus, for various reasons there may be a necessity to reduce the dimension m of the searching subspace before doing the next iteration step with the formulas (7.36-39).

This procedure was found to be very effective in self-consistent electronic structure calculations. The average speed of convergence was about one order of magnitude for Y^2 per iteration step so that generally the number of steps needed for convergency within the demanded accuracy was less than the dimension n of the problem, that is, less than one complete gradient to (7.34) had to be calculated. The procedure easily converged even in cases where the simple mixing procedure (7.32) ran into trouble.

These remarks close the detailed description of the ideas and the numerics of our version of LCAO electronic structure calculations. In Appendix 3. the reader will find the calculated scf-electronic structures of element metals. He may either take them as a further illustration of the power of the method or use the numbers for dealing with any problem where the detailed electronic structure is needed as an input.

APPENDIX 1. Spherical harmonics and their transformations

The spherical harmonics of various authors differ by constant phase factors. We use the definition of [85]:

$$\mathcal{Y}_{lm}(\zeta,\varphi) = i^{l} \sqrt{\frac{2l+1}{4}}\, \frac{(-1)^{(|m|+m)/2}\, e^{im\varphi}}{\sqrt{(l+|m|)!\,(l-|m|)!}}\, (1-\zeta^2)^{|m|/2}\, \zeta^{l-|m|}\, *$$

$$* \sum_{p=0,2,\dots}^{l-|m|} \frac{(-1)^{p/2}(l-|m|)!\,(2l-1-p)!!}{(l-|m|-p)!\; p!!}\, \zeta^{-p}\,, \qquad (A1.1)$$

$$\zeta = z/r\,, \quad \varphi = \mathrm{atan}(y/x)\,, \quad 0 \le \varphi \le 2\pi\,,$$

and define "real" spherical harmonics (in fact being real up to the factor i^{l} needed for the inversional symmetry)

$$Y_{lm} = \begin{cases} \dfrac{i}{\sqrt{2}}\left(\mathcal{Y}_{l-|m|} - (-1)^m \mathcal{Y}_{l|m|}\right) & < \\ \mathcal{Y}_{l0} & \text{for } m = 0\,. \\ \dfrac{1}{\sqrt{2}}\left(\mathcal{Y}_{l-|m|} + (-1)^m \mathcal{Y}_{l|m|}\right) & > \end{cases} \qquad (A1.2)$$

Both types of functions have the property

$$\mathcal{Y}_{lm}(-\mathbf{r}) = \mathcal{Y}^{*}_{lm}(\mathbf{r})\,, \quad Y_{lm}(-\mathbf{r}) = Y^{*}_{lm}(\mathbf{r}) \qquad (A1.3)$$

used in (2.56). Furthermore, the matrix elements (2.66) are the same independently wether the functions (A1.1) or (A1.2) are used. The first harmonics (A1.2) are

$$Y_{00} = 1/\sqrt{4\pi}\,,$$

$$Y_{1-1} = i\sqrt{\frac{3}{4\pi}}\,\frac{y}{r}\,, \quad Y_{10} = i\sqrt{\frac{3}{4\pi}}\,\frac{z}{r}\,, \quad Y_{11} = i\sqrt{\frac{3}{4\pi}}\,\frac{x}{r}\,,$$

$$Y_{2-2} = -\sqrt{\frac{15}{4\pi}}\,\frac{xy}{r^2}\,, \quad Y_{2-1} = -\sqrt{\frac{15}{4\pi}}\,\frac{yz}{r^2}\,, \quad Y_{20} = -\sqrt{\frac{5}{16\pi}}\left(3\,\frac{z^2}{r^2} - 1\right)\,,$$

$$Y_{21} = -\sqrt{\frac{15}{4\pi}}\,\frac{xz}{r^2}\ , \qquad Y_{22} = -\sqrt{\frac{15}{16\pi}}\,\frac{(x^2 - y^2)}{r^2}\ ,$$

$$Y_{3-3} = -i\sqrt{\frac{35}{32\pi}}\,\frac{3x^2y - y^3}{r^3}\ , \qquad Y_{3-2} = -i\sqrt{\frac{105}{4\pi}}\,\frac{xyz}{r^3}\ ,$$

$$Y_{3-1} = -i\sqrt{\frac{21}{32\pi}}\,\frac{y}{r}\,(5\,\frac{z^2}{r^2} - 1)\ , \qquad Y_{30} = -i\sqrt{\frac{7}{16\pi}}\,(5\,\frac{z^3}{r^3} - 3\,\frac{z}{r})\ ,$$

$$Y_{31} = -i\sqrt{\frac{21}{32\pi}}\,\frac{x}{r}\,(5\,\frac{z^2}{r^2} - 1)\ , \qquad Y_{32} = -i\sqrt{\frac{105}{16\pi}}\,\frac{(x^2 - y^2)z}{r^3}\ ,$$

$$Y_{33} = -i\sqrt{\frac{35}{32\pi}}\,\frac{x^3 - 3xy^2}{r^3}\ , \qquad \text{(A1.4)}$$

If $Y_{lm}(x,y,z)$ is a harmonic (A1.2) defined with respect to a given x-y-z-coordinate system, and $Y_{lm}(x',y',z')$ is another one with respect to a coordinate system the z'-axis of which has the direction of the unit vector (a,b,c) in the unprimed coordinates, then

$$Y_{lm}(x,y,z) = \sum_{m'} C^l_{mm'}\, Y_{lm'}(x',y',z')\ . \qquad \text{(A1.5)}$$

In applications of these transformation relations the actual orienation of the x'- and y'-axes is usually irrelevant. As the explicit expressions of the unitary transformation matrices C^l , however, depend on this orientation, it is arbitrarily fixed here so that the x-axis is in the x'-z'-plane, hence

$$\begin{aligned} x &= s\,x' + a\,z'\ , \\ y &= -\,a\frac{b}{s}\,x' + \frac{c}{s}\,y' + b\,z'\ , \qquad s \equiv \sqrt{1 - a^2}\ , \\ z &= -\,a\frac{c}{s}\,x' - \frac{b}{s}\,y' + c\,z'\ . \end{aligned} \qquad \text{(A1.6)}$$

Then the explicit transformation matrices are

$$C^0 = 1\ ,$$

$$C^1 = \begin{pmatrix} s & 0 & a \\ -ab/s & c/s & b \\ -ac/s & -b/s & c \end{pmatrix} ,$$

$$C^2 = \begin{pmatrix} c & ac/s & \sqrt{3}\ ab & b(1-2a^2)/s & -ab \\ a(b^2-c^2)/s^2 & (c^2-b^2)/s^2 & \sqrt{3}\ bc & -2abc/s & bc(1+a^2)/s^2 \\ \sqrt{3}\ abc/s^2 & -\sqrt{3}\ bc/s & \frac{1}{2}(3c^2-1) & -\sqrt{3}ac^2/s & \frac{\sqrt{3}}{2}(a^2c^2-b^2)/s^2 \\ -b & -ab/s & \sqrt{3}\ ac & c(1-2a^2)/s & -ac \\ abc/s^2 & -bc/s & \frac{\sqrt{3}}{2}(a^2-b^2) & a(2b^2+c^2)/s & \frac{1}{2}(1+c^2-2a^2b^2/s^2) \end{pmatrix}$$

$$C^3 = \left(\begin{matrix} -c(3(a^2-b^2)-4c^2/s^2)/4s & \sqrt{6}\ ac(2b^2+c^2)/2s^2 & \cdots \\ -\sqrt{6}\ a(b^2-c^2)/2s & -(b^2-c^2)(1-2a^2)/s^2 & \cdots \\ \sqrt{15}\ c(2b^2-c^2-(3b^2-c^2)/s^2)/4s & \sqrt{10}\ ac(2b^2-c^2)/2s^2 & \cdots \\ \sqrt{10}\ b(3c^2+(b^2-3c^2)/s^2)/4s & \sqrt{15}\ abc^2/s^2 & \cdots \\ \sqrt{15}\ abc/2s & -\sqrt{10}\ bc(1-2a^2)/2s^2 & \cdots \\ \sqrt{6}\ b(a^2-b^2+2c^2-4c^2/s^2)/4s & -ab(2b^2-c^2)/s^2 & \cdots \\ 3abc/2s & -\sqrt{6}\ bc(1-2a^2)/2s^2 & \cdots \end{matrix} \right.$$

$$\begin{matrix} \cdots & \sqrt{15}\ c(a^2-b^2)/4s & \sqrt{10}\ b(3a^2-b^2)/4 & \sqrt{15}\ ab(2-3a^2+b^2)/4s & \cdots \\ \cdots & -\sqrt{10}\ a(b^2-c^2)/2s & \sqrt{15}\ abc & \sqrt{10}\ bc(1-3a^2)/2s & \cdots \\ \cdots & -c(a^2+11b^2-4c^2)/4s & -\sqrt{6}\ b(1-5c^2)/4 & ab(1-15c^2)/4s & \cdots \\ \cdots & \sqrt{6}\ b(1-5c^2)/4s & -c(3-5c^2)/2 & \sqrt{6}\ ac(1-5c^2)/4s & \cdots \\ \cdots & -5abc/2s & -\sqrt{6}\ a(1-5c^2)/4 & -(b^2-4c^2+15a^2c^2)/4s & \cdots \\ \cdots & -\sqrt{10}\ b(a^2-b^2+2c^2)/4s & \sqrt{15}\ c(a^2-b^2)/2 & \sqrt{10}\ ac(2-3a^2+3b^2)/4s & \cdots \\ \cdots & \sqrt{15}\ abc/2s & \sqrt{10}\ a(a^2-3b^2)/4 & \sqrt{15}\ s(a^2-b^2+2a^2b^2/s^2)/4 & \cdots \end{matrix}$$

$$\left. \begin{matrix} \cdots & -\sqrt{6}\ b(3a^2-b^2-2c^2/s^2)/4 & ab(3a^2-b^2-(2b^2+6c^2)/s^2)/4s \\ \cdots & -abc(1-3a^2)/s^2 & \sqrt{6}\ bc(1+a^2)/2s \\ \cdots & -\sqrt{10}\ b(3c^2+(b^2-5c^2)/s^2)/4 & \sqrt{15}\ ab(c^2+(b^2-3c^2)/s^2)/4s \\ \cdots & -\sqrt{15}\ c(c^2+(b^2-c^2)/s^2)/2 & \sqrt{10}\ ac(c^2+(3b^2-c^2)/s^2)/4s \\ \cdots & -\sqrt{10}\ a(3c^2+(b^2-c^2)/s^2)/4 & \sqrt{15}\ (a^2c^2-b^2)/4s \\ \cdots & -c(3(a^2-b^2)/2+(2b^2-c^2)/s^2) & \sqrt{6}\ ac(a^2-b^2+2(b^2-c^2)/s^2)/4s \\ \cdots & -\sqrt{6}\ a(a^2-3b^2-2c^2/s^2)/4 & -s(a^2-b^2-(4c^2-2a^2b^2)/s^2)/4 \end{matrix} \right)$$

(A1.7)

The singular cases of these formulas are $a = \pm 1$, where instead of (A1.6) we keep the y-axis fixed:

$$\begin{aligned} x &= az' \\ y &= y' \\ z &= -ax' \end{aligned} \qquad \text{(A1.8)}$$

and have

$$C^0 = 1 ,$$

$$C^1 = \begin{pmatrix} 0 & 0 & a \\ 0 & 1 & 0 \\ -a & 0 & 0 \end{pmatrix},$$

$$C^2 = \begin{pmatrix} 0 & a & 0 & 0 & 0 \\ -a & 0 & 0 & 0 & 0 \\ 0 & 0 & -1/2 & 0 & \sqrt{3}/2 \\ 0 & 0 & 0 & -1 & 0 \\ 0 & 0 & \sqrt{3}/2 & 0 & 1/2 \end{pmatrix},$$

$$C^3 = \begin{pmatrix} 1/4 & 0 & \sqrt{15}/4 & 0 & 0 & 0 & 0 \\ 0 & -1 & 0 & 0 & 0 & 0 & 0 \\ \sqrt{15}/4 & 0 & -1/4 & 0 & 0 & 0 & 0 \\ 0 & 0 & 0 & 0 & (\sqrt{3/8})a & 0 & -(\sqrt{5/8})a \\ 0 & 0 & 0 & -(\sqrt{3/8})a & 0 & (\sqrt{5/8})a & 0 \\ 0 & 0 & 0 & 0 & -(\sqrt{5/8})a & 0 & -(\sqrt{3/8})a \\ 0 & 0 & 0 & (\sqrt{5/8})a & 0 & (\sqrt{3/8})a & 0 \end{pmatrix}. \tag{A1.9}$$

APPENDIX 2. Some useful theorems on basis function expansion

First we show that variational approximants to the eigenvalue of an Hermitean operator bounded from below estimate <u>all</u> the eigenvalues from above.

Let S be the eigenvector space of a Hermitean operator $\hat{h}$ bounded from below, that is, let S be spanned by the eigenvectors ψ_0, ψ_1, ψ_2, ... of $\hat{h}$ with real eigenvalues $\varepsilon_0 \leq \varepsilon_1 \leq \varepsilon_2 \leq \dots$ in ascending order. By $S_{\{\psi_0,\psi_1,\dots,\psi_n\}}$ we denote the subspace of S spanned by the vectors ψ_0, $\psi_1,\dots,\psi_n$, and by $S^{\perp}_{\{\psi_0,\psi_1,\dots,\psi_n\}}$ we denote the subspace of S consisting of all vectors being orthogonal to all vectors of $S_{\{\psi_0,\psi_1,\dots,\psi_n\}}$. Then

$$\varepsilon_0 = \min_{S} \{ \langle\psi|\hat{h}|\psi\rangle \mid \langle\psi|\psi\rangle = 1 \},$$

$$\varepsilon_1 = \min_{S^{\perp}_{\{\psi_0\}}} \{ \langle\psi|\hat{h}|\psi\rangle \mid \langle\psi|\psi\rangle = 1 \},$$

$$\varepsilon_2 = \min_{S^{\perp}_{\{\psi_0,\psi_1\}}} \{ \langle\psi|\hat{h}|\psi\rangle \mid \langle\psi|\psi\rangle = 1 \},$$

$$\dots . \tag{A2.1}$$

Let further $\bar{S}$ be a finite dimensional subspace of S spanned by some given basis, and let $\hat{P}$ be the orthogonal projector $\hat{P}S = \bar{S}$. What a finite dimensional variational solution of the eigenvalue problem for $\hat{h}$ yields is the eigenvectors $\varphi_0, \varphi_1, \varphi_2, \ldots$ and the eigenvalues $\eta_0 \leq \eta_1 \leq \eta_2 \leq \cdots$ of $\hat{P}\hat{h}\hat{P}$ in $\bar{S}$. For these quantities,

$$\eta_0 = \langle\varphi_0|\hat{h}|\varphi_0\rangle \;,\; \langle\varphi_0|\varphi_0\rangle = 1 \;,$$

$$\eta_1 = \max_{S_{\{\varphi_0,\varphi_1\}}} \{ \langle\psi|\hat{h}|\psi\rangle \mid \langle\psi|\psi\rangle = 1 \} \;,$$

$$\eta_2 = \max_{S_{\{\varphi_0,\varphi_1,\varphi_2\}}} \{ \langle\psi|\hat{h}|\psi\rangle \mid \langle\psi|\psi\rangle = 1 \} \;,$$

$$\ldots \; . \tag{A2.2}$$

From the first relations (A2.1) and (A2.2), since $\varphi_0 \in \bar{S} \subset S$, obviously

$$\varepsilon_0 \leq \eta_0 \; . \tag{A2.3}$$

Let further

$$\chi_1 \in S_{\{\varphi_0,\varphi_1\}} \cap S^{\perp}_{\{\psi_0\}} \neq 0, \quad \langle\chi_1|\chi_1\rangle = 1 \; . \tag{A2.4}$$

Such a vector exists because in any two-dimensional subspace of S (as $S_{\{\varphi_0,\varphi_1\}}$) a vector orthogonal to any given one-dimensional subspace of S (as $S_{\{\psi_0\}}$) can be found. The second relations of (A2.1) and (A2.2) now yield

$$\varepsilon_1 \leq \langle\chi_1|\hat{h}|\chi_1\rangle \leq \eta_1 \; . \tag{A2.5}$$

Next use

$$\chi_2 \in S_{\{\varphi_0,\varphi_1,\varphi_2\}} \cap S^{\perp}_{\{\psi_0,\psi_1\}} \neq 0 \;, \quad \langle\chi_2|\chi_2\rangle = 1 \;, \tag{A2.6}$$

existing by an analogous argument, and so on. This proves the theorem

$$\boxed{\varepsilon_i \leq \eta_i \;, \quad i = 0,1,2,\ldots \; .} \tag{A2.7}$$

(Replacing $\hat{h}$ by $-\hat{h}$ yields the analogous theorem with all inequality signs reversed for an operator bounded from above.)

Let now $v \in S$ be any given vector, and let $\bar{v} \in \bar{S}$ such that

$$\langle v - \bar{v} | v - \bar{v} \rangle \Longrightarrow \min. \tag{A2.8}$$

Then, clearly from geometrical reasoning,

$$(v - \bar{v}) \in \bar{S}^{\perp} . \tag{A2.9}$$

Formally this is obtained by introducing a (finite) orthonormal basis $\{b_i\}$ in $\bar{S}$ and putting

$$v - \bar{v} = \sum_i c_i b_i + \delta v , \quad \langle b_i | \delta v \rangle = 0 \text{ for all } i.$$

Then, from (A2.8),

$$0 = \frac{\delta}{\delta c_i} \langle v - \bar{v} | v - \bar{v} \rangle = \frac{\delta}{\delta c_i} \Big(\sum_j c_j^2 + \langle \delta v | \delta v \rangle \Big) = 2c_i ,$$

and hence $v - \bar{v} = \delta v \in \bar{S}^{\perp}$. For any vector $\bar{\varphi} \in \bar{S}$, (A2.9) implies

$$\boxed{\langle v - \bar{v} | \varphi \rangle = \langle v - \bar{v} | \varphi - \bar{\varphi} \rangle ,} \tag{A2.10}$$

being just the theorem stated on page 37 in section 2.1.

APPENDIX 3. Results of DFT-LCAO bandstructure calculations for elemental metals

This appendix contains the complete results of self-consistent LCAO bandstructure calculations using the Hedin-Lundqvist exchange and correlation potential expression (1.34-35) for all metals up to the atomic number 30 (Zn). The lattice constants corresponding to T = 0K were taken from experiment [87], and the observed low-temperature structure was used except for the light alkaline metals which were treated in the b.c.c. structure instead of h.c.p., and for Manganese being treated in the f.c.c. structure with a lattice constant corresponding to the experimental density. Atom-like basis functions were calculated from (3.2) with x_l-values (see (3.6) for the definition) $x_0 = x_2 = 1.$, $x_1 = 0.925$. The only exception is vanadium, where x_0 had to be reduced to $x_0 = 0.925$ for the sake of numerical stability (see the discussion on page 56). For each metal, the tables contain in turn the exponents α' and the coefficients $v_{\nu\alpha}$ (the column under each α-value corresponds to $\nu = 1,2,3$ in turn) of a representation (2.20) of the self-consistent site potential entering (7.3), the exponents α and the coefficients $a_{\nu\alpha}^{(n)}$, n = 1s, 2s, 2p, ... , of a representation (2.26), (2.6) of the atom-like basis functions obtained from (3.1) and entering (2.38),(2.50-51), the exponents α_t and the coefficients $\varrho_t^{(S)}$ of a representation (5.29), (5.34) of the total site density (including the core-electron density) entering (5.32), and (with the same exponents) the coefficients $\tilde{\varrho}_t^{(S)}$ of a representation (5.46) of the total crystal electron density, spherically averaged within an atomic sphere the radius $r_{at}^{(S)}$ of which is given by (7.22). Next the Slater-Koster matrix elements are given in the same arrangement as in Table 4.1 and as explained in section 4.1. The neighbouring distances are in turn:

	neighbour	distance
f.c.c.	(110)	$a/\sqrt{2}$
	(200)	a
	(211)	$a\sqrt{3/2}$
	(220)	$a\sqrt{2}$
b.c.c.	(111)	$a\sqrt{3/4}$
	(200)	a
	(220)	$a\sqrt{2}$
	(311)	$a\sqrt{11/4}$
	(222)	2a
h.c.p.	(0a0)	a
	(a/2 3,a/2,c/2)	$\sqrt{a^2/3+c^2/4}$
	(2a/ 3,0,c/2)	$\sqrt{4a^2/3+c^2/4}$
	(00c)	c
	(3a,0,0)	$\sqrt{3}a$
	(2a/ 3,a,c/2)	$\sqrt{7a^2/3+c^2/4}$
	(0ac)	$\sqrt{a^2+c^2}$
	(0,2a,0)	2a

Finally, the Fermi energy is given on the scale defined by the potential representation (7.3).

The tables are followed by a plot of $z_S(r) = -r \cdot v_S(r)$, a bandstructure plot on all symmetry lines and a density-of-states plot. The bandstructure may easily be reproduced from the table of Slater-Koster matrix elements. The density of states and the Fermi energy were obtained by the simplex method of chapter 6. using 89 **k**-points for the f.c.c. structure, 91 for b.c.c., and 76 for h.c.p.

```
LITHIUM, B.C.C. STRUCTURE, A = 6.59 A.U.

 EXPONENTS FOR THE POTENTIAL BASIS:
  3.0000      1.3155
 POTENTIAL EXPANSION COEFFICIENTS:
  0.32675E+01-0.26754E+00
  0.29190E+01 0.22423E+01
  0.10996E+01 0.60701E-01

 EXPONENTS FOR THE ATOM-LIKE ORBITAL BASIS:
  3.0000      1.3155
 ORBITAL BASIS EXPANSION COEFFICIENTS:
   1S:
 -0.93195E+01 0.26018E+00
 -0.47247E+00-0.23248E+00
 -0.31712E+01 0.36007E-01
   2S:
 -0.18688E+02 0.20975E+02
 -0.22767E+02-0.10969E+02
 -0.16399E+02 0.30085E+00
   2P:
 -0.10484E+02 0.91828E+01
 -0.97917E+01-0.58187E+01
 -0.82107E+01 0.62230E+00

 EXPONENTS FOR THE DENSITY BASIS:
  6.0000      2.8094      1.3155
 SITE DENSITY EXPANSION COEFFICIENTS:
  0.17033E+03-0.77636E+01 0.21562E+01
  0.17146E+02 0.13439E+02 0.44602E+00
  0.81430E+02-0.14523E+02-0.48994E-01
 TOTAL ATOMIC SPHERE DENSITY EXPANSION COEFFICIENTS:
  0.21563E+03-0.21505E+03 0.16286E+03
  0.86754E+02-0.55344E+02-0.79731E+02
  0.62898E+02-0.15083E+03 0.11187E+02
```

SLATER-KOSTER INTEGRALS:

HAMILTONIAN INTEGRALS:

ONE-CENTRE:

(S)	(PO)	(P1-)	(P1+)
0.0878	0.1937	0.1937	0.1937

TWO-CENTRE:

(SSS)	(PSS)	(PPS)	(PPP)
-0.0175	-0.0105	-0.0063	-0.0089
-0.0176	-0.0178	0.0140	-0.0068
-0.0045	-0.0056	0.0078	-0.0006
-0.0012	-0.0012	0.0014	-0.0000
-0.0005	-0.0002	-0.0000	-0.0000

OVERLAP INTEGRALS:

ONE-CENTRE:

(S)	(PO)	(P1-)	(P1+)
0.9952	0.9975	0.9975	0.9975

TWO-CENTRE:

(SSS)	(PSS)	(PPS)	(PPP)
0.2081	0.2344	-0.3084	0.0761
0.1227	0.1398	-0.1919	0.0317
0.0150	0.0120	-0.0122	-0.0002
0.0035	0.0009	0.0015	-0.0005
0.0019	-0.0003	0.0027	-0.0004

FERMI ENERGY = 0.0608 HARTREE

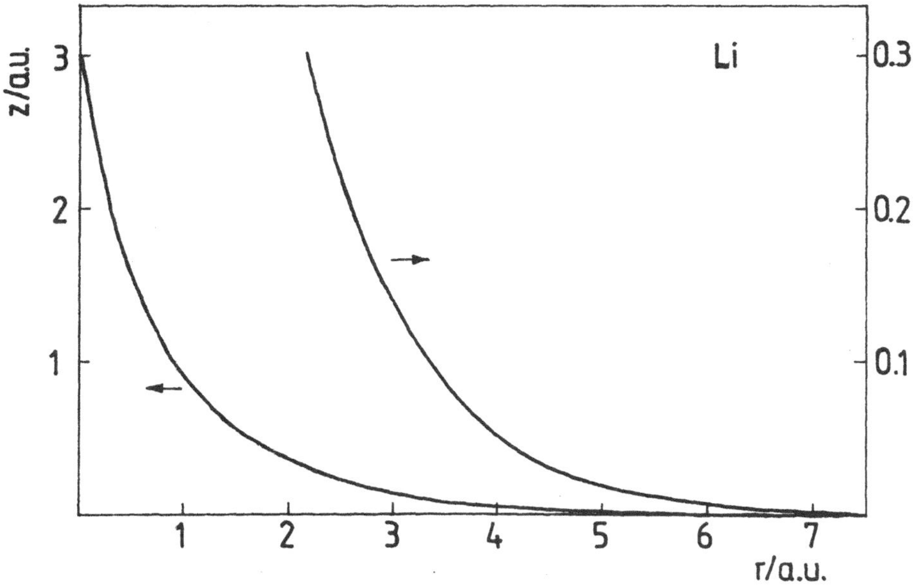

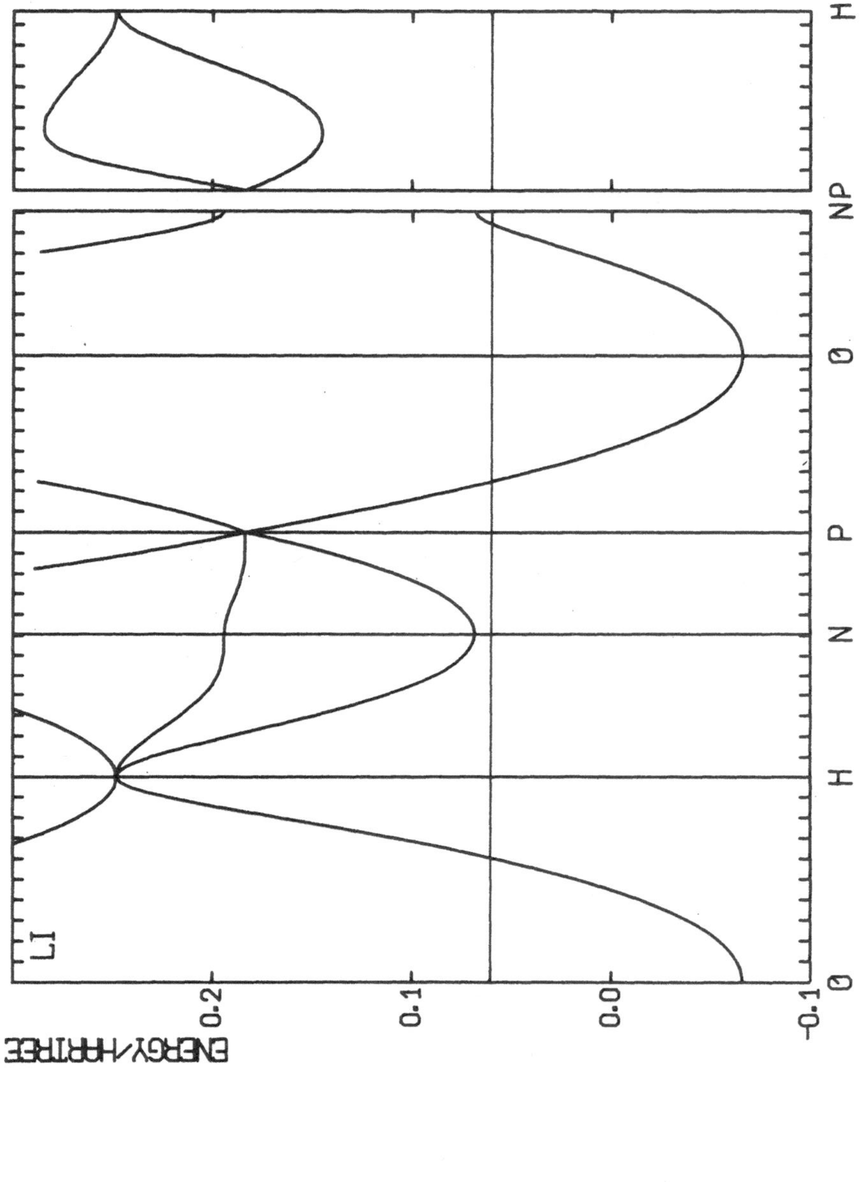
LI
ENERGY/HARTREE
0.2
0.1
0.0
-0.1
Ø
H
N
P
Ø
NP
H

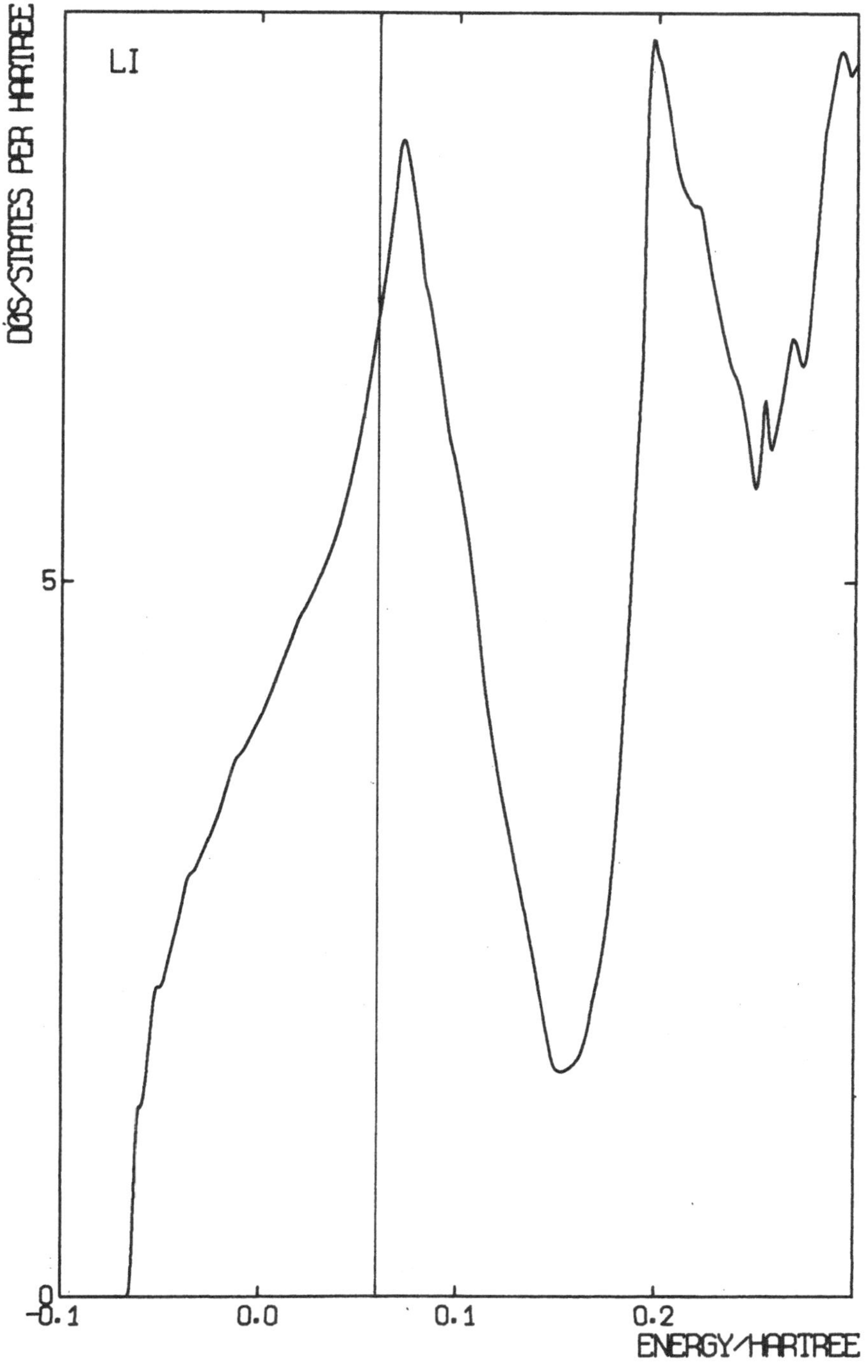
LI
DOS/STATES PER HARTREE
5
0
-0.1
0.0
0.1
0.2
ENERGY/HARTREE

```
BERYLLIUM, H.C.P. STRUCTURE, A = 4.32 A.U., C = 6.76 A.U.

 EXPONENTS FOR THE POTENTIAL BASIS:
  4.0000       1.7931
 POTENTIAL EXPANSION COEFFICIENTS:
  0.76854E+01-0.36854E+01
  0.89983E+01 0.87792E+01
  0.19724E+01-0.68747E+00

 EXPONENTS FOR THE ATOM-LIKE ORBITAL BASIS:
  4.0000       1.7931
 ORBITAL BASIS EXPANSION COEFFICIENTS:
   1S:
  0.15064E+02-0.63253E+00
  0.11336E+01 0.50838E+00
  0.78536E+01-0.92304E-01
   2S:
 -0.37418E+02 0.41273E+02
 -0.57312E+02-0.30721E+02
 -0.49521E+02 0.26414E+01
   2P:
 -0.17697E+02 0.14730E+02
 -0.22464E+02-0.13964E+02
 -0.19708E+02 0.21256E+01

 EXPONENTS FOR THE DENSITY BASIS:
  8.0000       3.7875       1.7931
 SITE DENSITY EXPANSION COEFFICIENTS:
  0.39747E+03-0.82011E+01 0.29572E+02
 -0.42342E+02-0.34584E+02-0.86216E+01
  0.28377E+02-0.93303E+02 0.10946E+01
 TOTAL ATOMIC SPHERE DENSITY EXPANSION COEFFICIENTS:
  0.41313E+02-0.27373E+03 0.63819E+03
 -0.90982E+03-0.12144E+04-0.44445E+03
 -0.20330E+04-0.50745E+03 0.88560E+02
```

SLATER-KOSTER INTEGRALS:

HAMILTONIAN INTEGRALS:

ONE-CENTRE:

(S)	(PO)	(P1-)	(P1+)
0.0949	0.2871	0.2918	0.2918

TWO-CENTRE:

(SSS)	(PSS)	(PPS)	(PPP)
-0.0474	-0.0437	0.0318	-0.0192
-0.0487	-0.0430	0.0278	-0.0206
-0.0177	-0.0216	0.0278	-0.0033
-0.0082	-0.0098	0.0132	-0.0008
-0.0033	-0.0036	0.0046	-0.0001
-0.0036	-0.0039	0.0050	-0.0001
-0.0014	-0.0012	0.0012	0.0000
-0.0005	-0.0000	-0.0004	0.0000

OVERLAP INTEGRALS:

ONE-CENTRE:

(S)	(PO)	(P1-)	(P1+)
0.9966	0.9980	0.9984	0.9984

TWO-CENTRE:

(SSS)	(PSS)	(PPS)	(PPP)
0.1626	0.1848	-0.2435	0.0510
0.1794	0.2031	-0.2651	0.0597
0.0285	0.0284	-0.0367	0.0020
0.0112	0.0086	-0.0088	-0.0004
0.0043	0.0017	0.0003	-0.0006
0.0047	0.0020	-0.0001	-0.0006
0.0019	-0.0002	0.0022	-0.0004
0.0007	-0.0007	0.0021	-0.0003

FERMI ENERGY = 0.2004 HARTREE

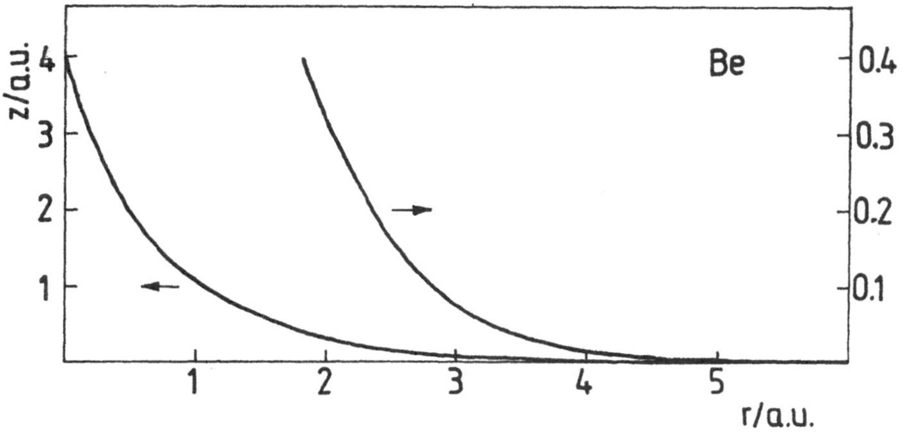

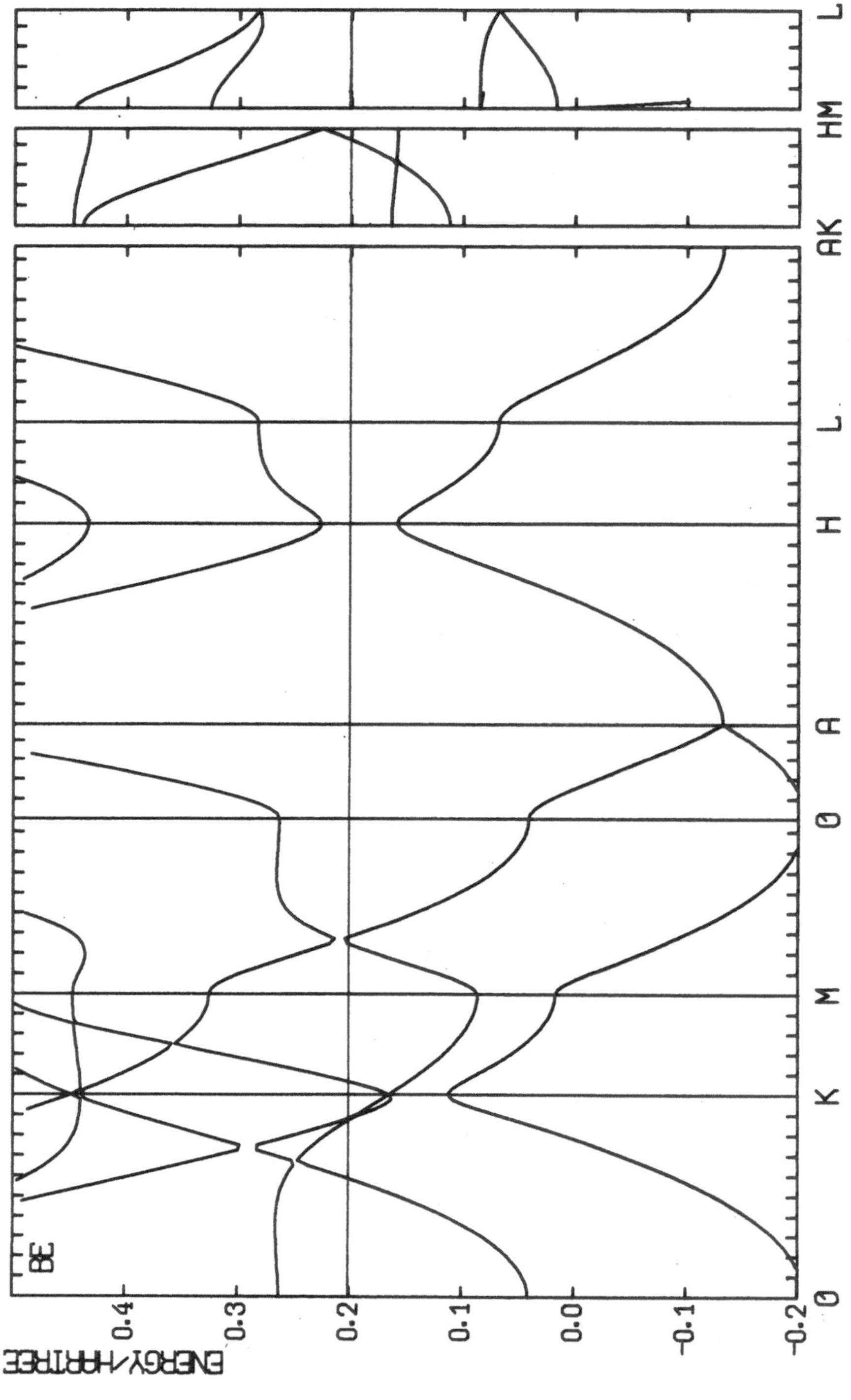
BE
ENERGY/HARTREE
0.4
0.3
0.2
0.1
0.0
-0.1
-0.2
0
K
M
0
A
H
L
AK
HM
L

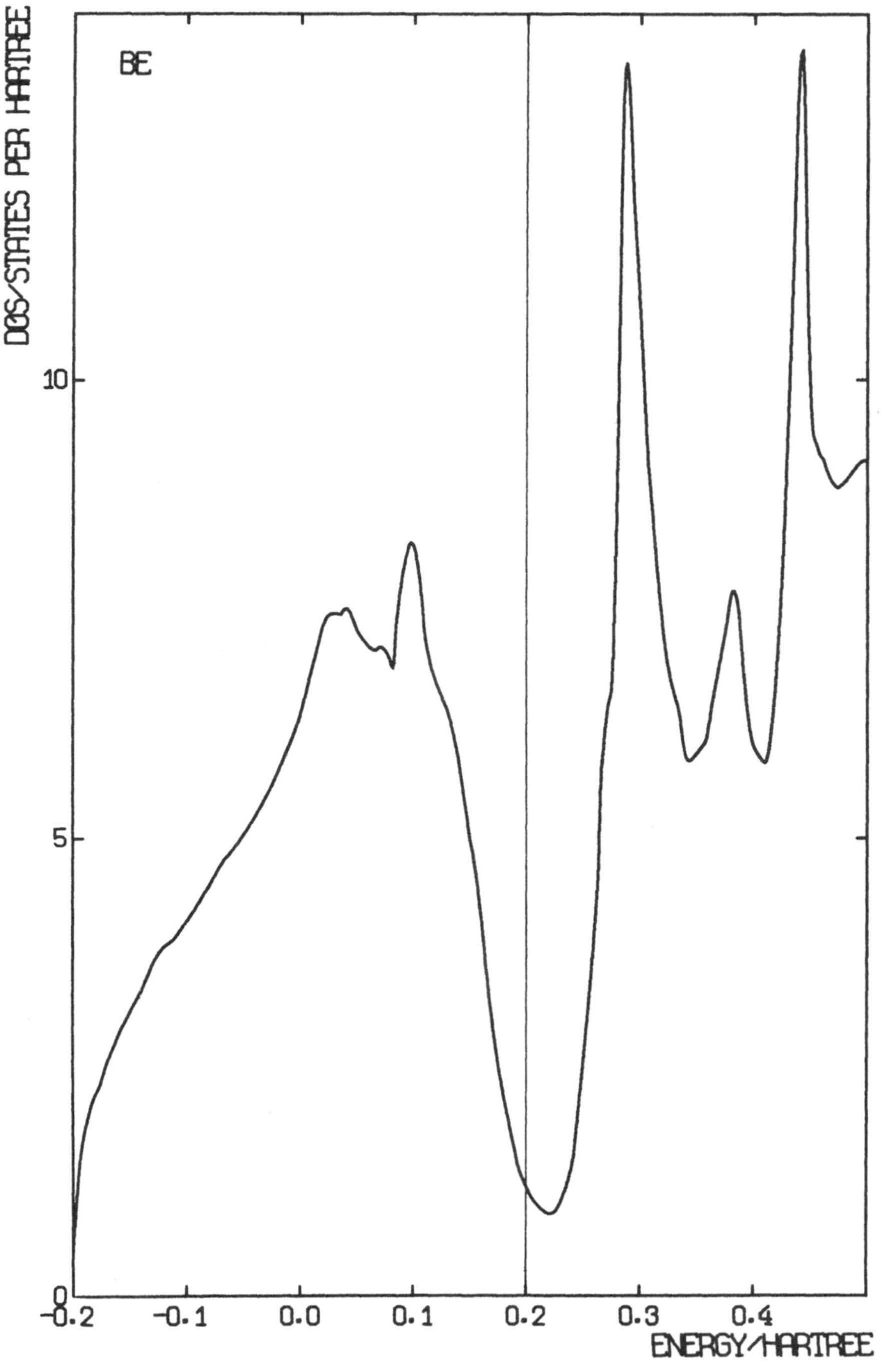
BE
DOS/STATES PER HARTREE
10
5
0
-0.2
-0.1
0.0
0.1
0.2
0.3
0.4
ENERGY/HARTREE

```
SODIUM, B.C.C. STRUCTURE, A = 7.79 A.U.

 EXPONENTS FOR THE POTENTIAL BASIS:
 11.0000      3.5069      1.1180
 POTENTIAL EXPANSION COEFFICIENTS:
 -0.74816E+01 0.15618E+02 0.28637E+01
 -0.41454E+02-0.13566E+02 0.10976E+01
 -0.11860E+03 0.14771E+02 0.31922E-02

 EXPONENTS FOR THE ATOM-LIKE ORBITAL BASIS:
 11.0000      3.5069      1.1180
 ORBITAL BASIS EXPANSION COEFFICIENTS:
   1S:
  0.70840E+02-0.14343E+01 0.11681E+00
  0.93161E+01 0.25553E+01-0.54933E-01
  0.14800E+03-0.18649E+01 0.62916E-02
   2S:
 -0.21376E+02 0.30105E+01 0.12076E+01
 -0.48277E+02 0.15236E+02-0.48427E+00
 -0.10261E+03 0.18602E+01 0.49121E-01
   2P:
  0.13546E+02 0.25952E+02 0.43783E+00
  0.27855E+02-0.13083E+02-0.13061E+00
  0.10079E+03 0.99671E+01 0.98220E-02
   3S:
  0.51478E+01-0.15702E+02 0.64857E+01
  0.32041E+02 0.21802E+02-0.36291E+01
  0.10692E+03-0.35011E+02-0.14861E+00
   3P:
  0.21288E+02-0.13940E+02 0.49423E+01
  0.52206E+02 0.26334E+02-0.30407E+01
  0.40130E+03-0.34217E+02 0.26696E+00

 EXPONENTS FOR THE DENSITY BASIS:
 22.0000      8.1488      3.0183      1.1180
 SITE DENSITY EXPANSION COEFFICIENTS:
  0.10565E+05-0.53312E+03 0.19944E+03-0.26167E+01
  0.46330E+04 0.56945E+03-0.12744E+03 0.15116E+01
  0.36578E+05 0.23819E+04 0.28474E+02-0.12738E+00
 TOTAL ATOMIC SPHERE DENSITY EXPANSION COEFFICIENTS:
  0.10217E+05-0.55041E+02-0.18777E+02 0.47294E+02
  0.30558E+04-0.32955E+03 0.78409E+02-0.24297E+02
  0.14252E+05 0.56605E+04-0.16131E+03 0.34953E+01
```

SLATER-KOSTER INTEGRALS:

HAMILTONIAN INTEGRALS:

ONE-CENTRE:

(S)	(P0)	(P1-)	(P1+)
0.0473	0.1594	0.1594	0.1594

TWO-CENTRE:

(SSS)	(PSS)	(PPS)	(PPP)
-0.0166	-0.0070	-0.0243	-0.0048
-0.0144	-0.0136	0.0027	-0.0055
-0.0036	-0.0050	0.0082	-0.0010
-0.0011	-0.0014	0.0022	-0.0002
-0.0008	-0.0010	0.0016	-0.0001

OVERLAP INTEGRALS:

ONE-CENTRE:

(S)	(P0)	(P1-)	(P1+)
0.9924	0.9921	0.9921	0.9921

TWO-CENTRE:

(SSS)	(PSS)	(PPS)	(PPP)
0.1984	0.2612	-0.3649	0.1026
0.1162	0.1593	-0.2461	0.0471
0.0150	0.0172	-0.0250	0.0017
0.0038	0.0030	-0.0022	-0.0001
0.0025	0.0018	-0.0008	-0.0002

FERMI ENERGY = 0.0573 HARTREE

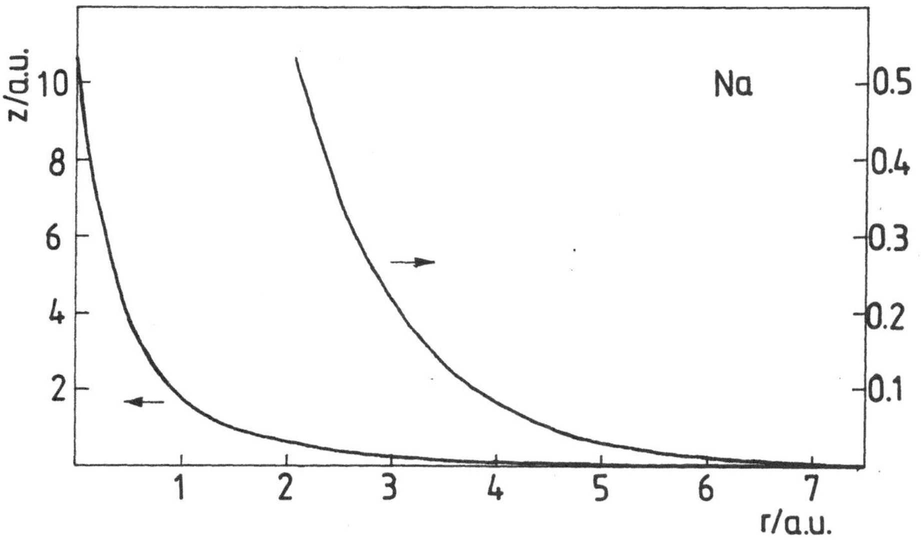

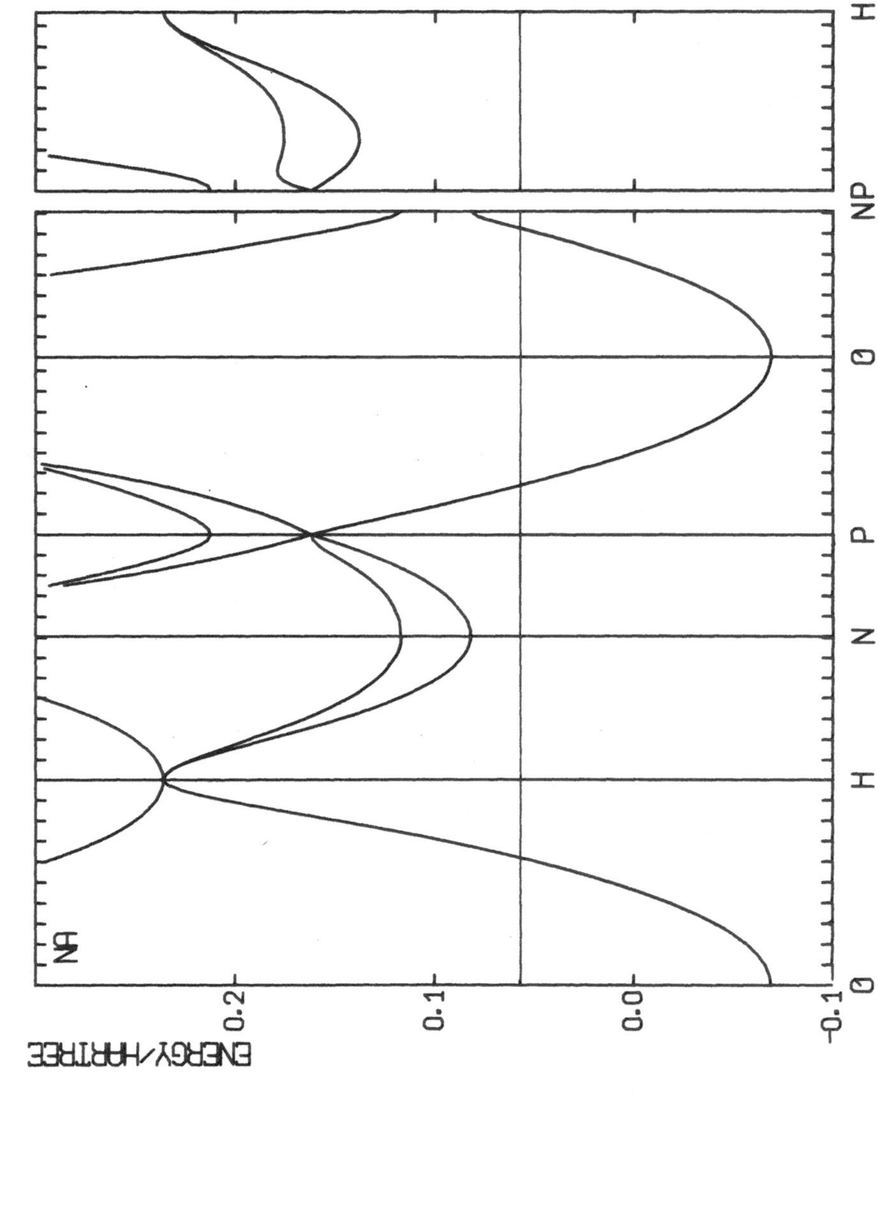
NA
ENERGY/HARTREE
0.2
0.1
0.0
-0.1
H
NP
H
P
N

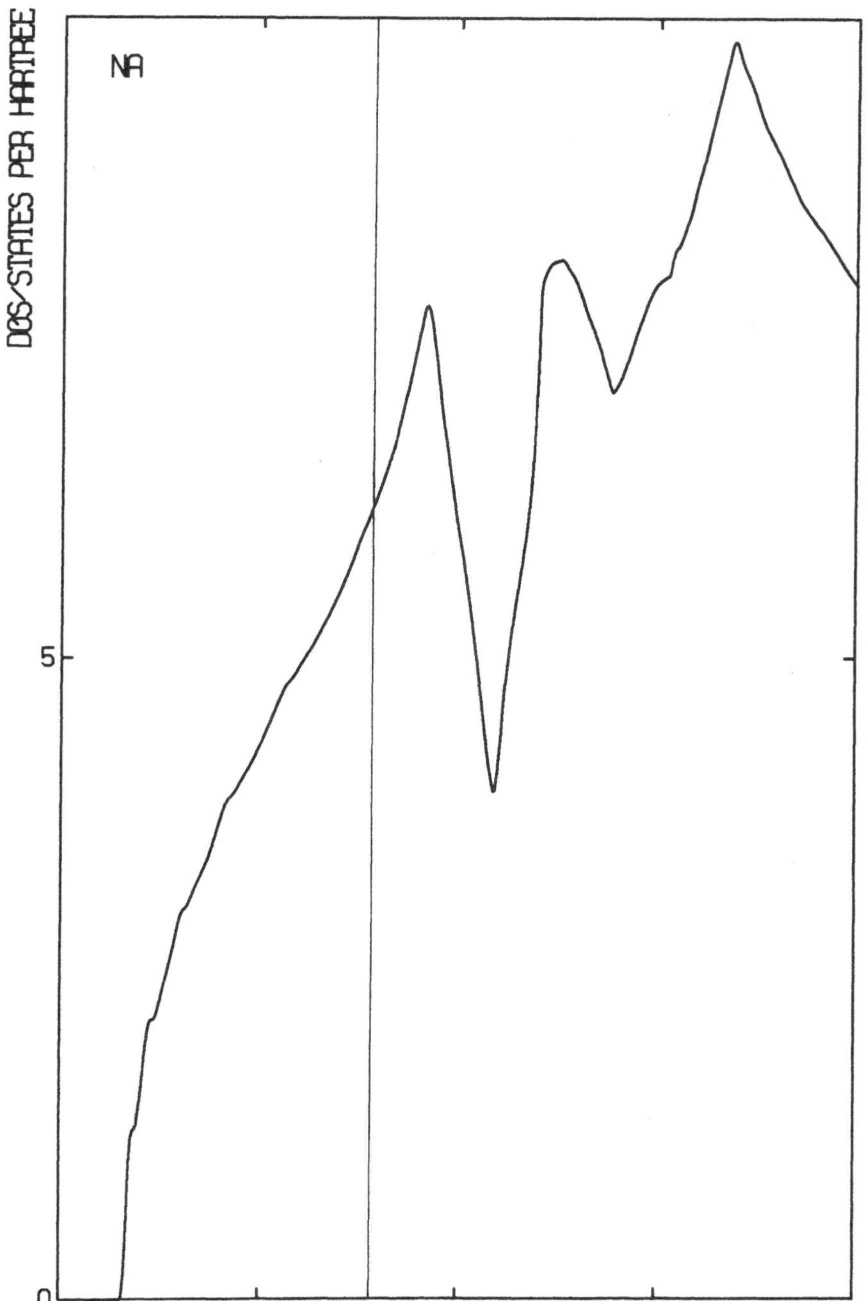
DOS/STATES PER HARTREE
NA
5
0
-0.1
0.0
0.1
0.2
ENERGY/HARTREE

MAGNESIUM, H.C.P. STRUCTURE, A = 6.06 A.U., C = 9.84 A.U.

```
 EXPONENTS FOR THE POTENTIAL BASIS:
 12.0000      3.9161      1.2780
 POTENTIAL EXPANSION COEFFICIENTS:
 -0.12270E+02 0.28663E+02-0.43934E+01
 -0.65638E+02-0.16432E+02 0.79375E+01
 -0.17388E+03 0.52106E+02-0.84856E+00

 EXPONENTS FOR THE ATOM-LIKE ORBITAL BASIS:
 12.0000      3.9161      1.2780
 ORBITAL BASIS EXPANSION COEFFICIENTS:
   1S:
  0.81090E+02-0.17345E+01 0.14531E+00
  0.11648E+02 0.33673E+01-0.76416E-01
  0.18973E+03-0.27322E+01 0.98148E-02
   2S:
  0.24588E+02-0.31285E+01-0.12768E+01
  0.60012E+02-0.22626E+02 0.58413E+00
  0.13015E+03-0.22423E+01-0.67308E-01
   2P:
  0.17632E+02 0.35111E+02 0.40785E+00
  0.41626E+02-0.18004E+02-0.14371E+00
  0.14592E+03 0.15367E+02 0.12663E-01
   3S:
  0.64422E+01-0.21999E+02 0.10179E+02
  0.43416E+02 0.28523E+02-0.72070E+01
  0.14903E+03-0.56319E+02 0.25508E+00
   3P:
  0.22976E+02-0.12499E+02 0.58170E+01
  0.66429E+02 0.28085E+02-0.44172E+01
  0.46227E+03-0.44239E+02 0.46971E+00

 EXPONENTS FOR THE DENSITY BASIS:
 24.0000      9.0293      3.3970      1.2780
 SITE DENSITY EXPANSION COEFFICIENTS:
  0.13495E+05-0.29972E+03 0.19019E+03 0.26282E+01
  0.47968E+04-0.66843E+03-0.79027E+02 0.81731E+00
  0.20509E+05 0.84959E+04-0.36036E+02-0.94894E-01
 TOTAL ATOMIC SPHERE DENSITY EXPANSION COEFFICIENTS:
  0.12586E+05 0.10199E+04-0.46469E+03 0.15939E+03
 -0.10413E+03-0.31284E+04 0.58974E+03-0.90240E+02
 -0.44065E+05 0.19411E+05-0.75016E+03 0.14237E+02
```

SLATER-KOSTER INTEGRALS:

HAMILTONIAN INTEGRALS:

ONE-CENTRE:

(S)	(P0)	(P1-)	(P1+)
-0.0034	0.1467	0.1473	0.1473

TWO-CENTRE:

(SSS)	(PSS)	(PPS)	(PPP)
-0.0323	-0.0301	0.0072	-0.0120
-0.0326	-0.0302	0.0067	-0.0120
-0.0091	-0.0130	0.0195	-0.0029
-0.0036	-0.0052	0.0088	-0.0009
-0.0020	-0.0028	0.0049	-0.0004
-0.0021	-0.0028	0.0049	-0.0004
-0.0008	-0.0010	0.0017	-0.0001
-0.0005	-0.0006	0.0009	-0.0000

OVERLAP INTEGRALS:

ONE-CENTRE:

(S)	(P0)	(P1-)	(P1+)
0.9954	0.9947	0.9949	0.9949

TWO-CENTRE:

(SSS)	(PSS)	(PPS)	(PPP)
0.1511	0.2113	-0.3144	0.0752
0.1532	0.2140	-0.3174	0.0767
0.0245	0.0327	-0.0558	0.0048
0.0080	0.0085	-0.0124	0.0003
0.0045	0.0038	-0.0041	-0.0002
0.0045	0.0038	-0.0043	-0.0002
0.0017	0.0007	0.0006	-0.0003
0.0010	0.0002	0.0011	-0.0003

FERMI ENERGY = 0.0982 HARTREE

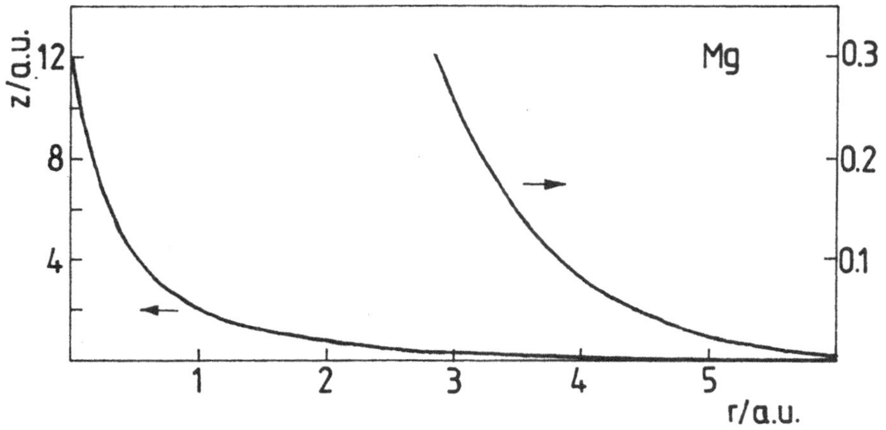

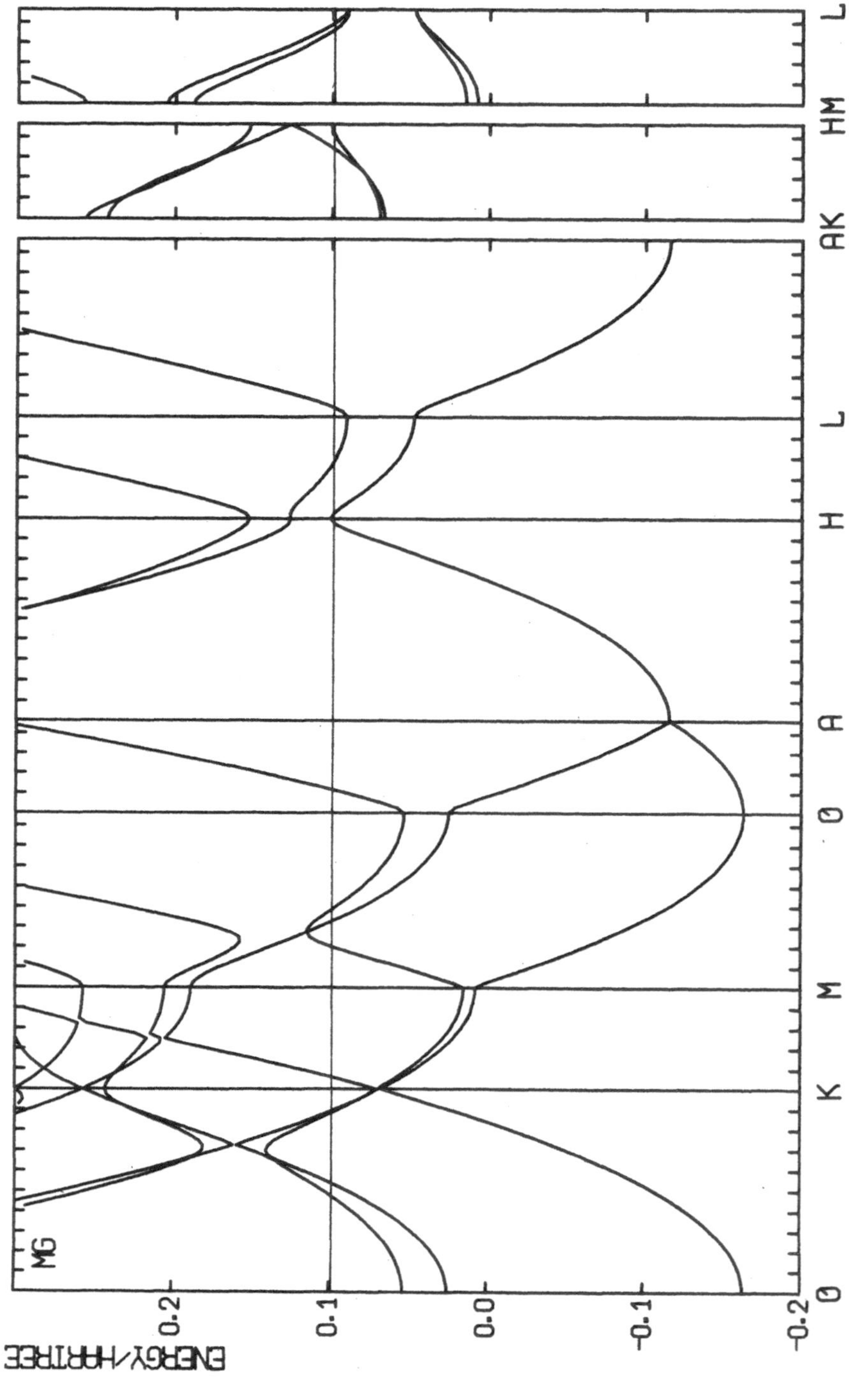
MG
ENERGY/HARTREE
0.2
0.1
0.0
-0.1
-0.2
K
M
A
H
L
AK
HM
L

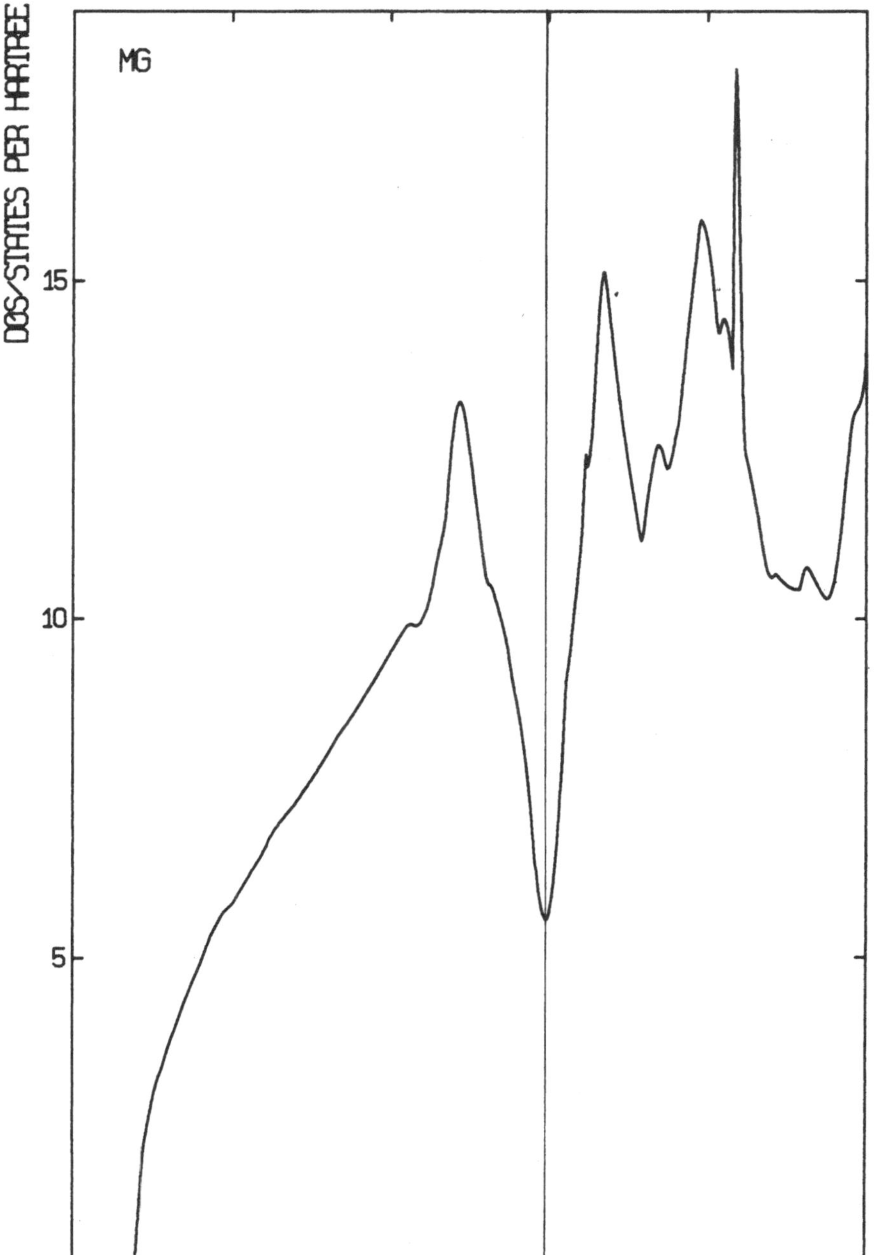
MG
DOS/STATES PER HARTREE
15
10
5
0
-0.2
-0.1
0.0
0.1
0.2
ENERGY/HARTREE

```
ALUMINIUM, F.C.C. STRUCTURE, A = 7.59 A.U.

 EXPONENTS FOR THE POTENTIAL BASIS:
 13.0000       4.3192        1.4350
 POTENTIAL EXPANSION COEFFICIENTS:
 -0.25594E+02 0.65608E+02-0.27014E+02
 -0.13472E+03-0.20200E+02 0.29738E+02
 -0.34092E+03 0.17048E+03-0.53123E+01

 EXPONENTS FOR THE ATOM-LIKE ORBITAL BASIS:
 13.0000       4.3192        1.4350
 ORBITAL BASIS EXPANSION COEFFICIENTS:
    1S:
  0.91879E+02-0.21062E+01 0.17873E+00
  0.14010E+02 0.44188E+01-0.10369E+00
  0.24228E+03-0.39265E+01 0.14724E-01
    2S:
  0.27877E+02-0.31773E+01-0.12794E+01
  0.72884E+02-0.32299E+02 0.65645E+00
  0.16128E+03-0.20583E+01-0.84533E-01
    2P:
  0.22743E+02 0.45621E+02 0.34568E+00
  0.60840E+02-0.23385E+02-0.13852E+00
  0.21837E+03 0.21649E+02 0.13824E-01
    3S:
  0.59389E+01-0.25542E+02 0.12759E+02
  0.47387E+02 0.33608E+02-0.10878E+02
  0.17255E+03-0.74110E+02 0.81211E+00
    3P:
  0.22727E+02-0.71088E+01 0.58778E+01
  0.74319E+02 0.24749E+02-0.56557E+01
  0.47485E+03-0.48336E+02 0.70851E+00
    3D:
 -0.31443E+01-0.37340E+01 0.10057E+01
  0.34974E+02-0.48354E+00-0.14941E+01
 -0.14721E+03-0.10428E+02 0.19642E+00

 EXPONENTS FOR THE DENSITY BASIS:
 26.0000       9.8992        3.7690        1.4350
 SITE DENSITY EXPANSION COEFFICIENTS:
  0.17960E+05-0.10864E+04 0.43609E+03-0.14874E+02
  0.87320E+04 0.10392E+04-0.38153E+03 0.12797E+02
  0.85117E+05 0.90227E+04 0.11240E+03-0.16608E+01
 TOTAL ATOMIC SPHERE DENSITY EXPANSION COEFFICIENTS:
  0.16862E+05 0.75672E+03-0.70744E+03 0.30440E+03
  0.14821E+04-0.16610E+04 0.78653E+03-0.19624E+03
  0.35033E+04 0.27281E+05-0.14694E+04 0.35583E+02
```

SLATER-KOSTER INTEGRALS:

HAMILTONIAN INTEGRALS:

ONE-CENTRE:

(S)	(P0)	(P1-)	(P1+)	(D0)	(D1-)	(D1+)	(D2-)	(D2+)
-0.0206	0.1780	0.1780	0.1780	0.4254	0.4005	0.4005	0.4005	0.4254

TWO-CENTRE:

(SSS)	(PSS)	(DSS)	(PPS)	(PPP)	(DPS)	(DPP)	(DDS)	(DDP)	(DDD)
-0.0436	-0.0451	-0.0158	0.0247	-0.0168	-0.0340	-0.0125	0.0833	-0.0277	-0.0020
-0.0099	-0.0144	-0.0157	0.0224	-0.0032	0.0203	-0.0070	-0.0067	0.0117	-0.0019
-0.0020	-0.0027	-0.0039	0.0050	-0.0004	0.0078	-0.0012	-0.0102	0.0035	-0.0003
-0.0005	-0.0005	-0.0006	0.0008	0.0000	0.0017	-0.0001	-0.0036	0.0006	-0.0000

OVERLAP INTEGRALS:

ONE-CENTRE:

(S)	(P0)	(P1-)	(P1+)	(D0)	(D1-)	(D1+)	(D2-)	(D2+)
0.9970	0.9964	0.9964	0.9964	0.9955	0.9909	0.9909	0.9909	0.9955

TWO-CENTRE:

(SSS)	(PSS)	(DSS)	(PPS)	(PPP)	(DPS)	(DPP)	(DDS)	(DDP)	(DDD)
0.1297	0.1880	0.1974	-0.2862	0.0645	-0.2748	0.1359	0.2100	-0.2457	0.0521
0.0179	0.0246	0.0332	-0.0442	0.0032	-0.0689	0.0100	0.0977	-0.0330	0.0020
0.0028	0.0020	0.0014	-0.0021	-0.0004	-0.0058	-0.0009	0.0161	-0.0005	-0.0005
0.0005	-0.0003	-0.0016	0.0015	-0.0003	0.0023	-0.0009	-0.0011	0.0018	-0.0003

FERMI ENERGY = 0.1711 HARTREE

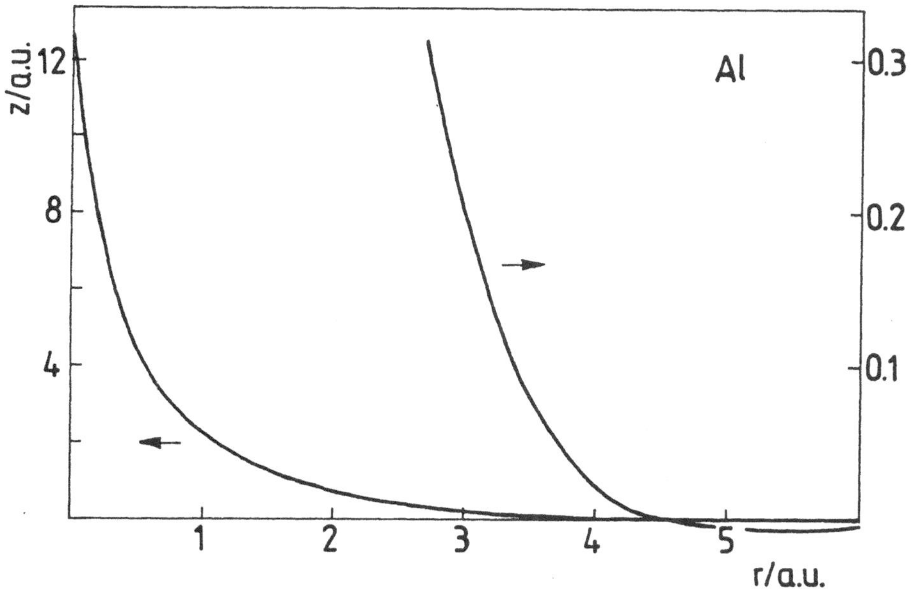

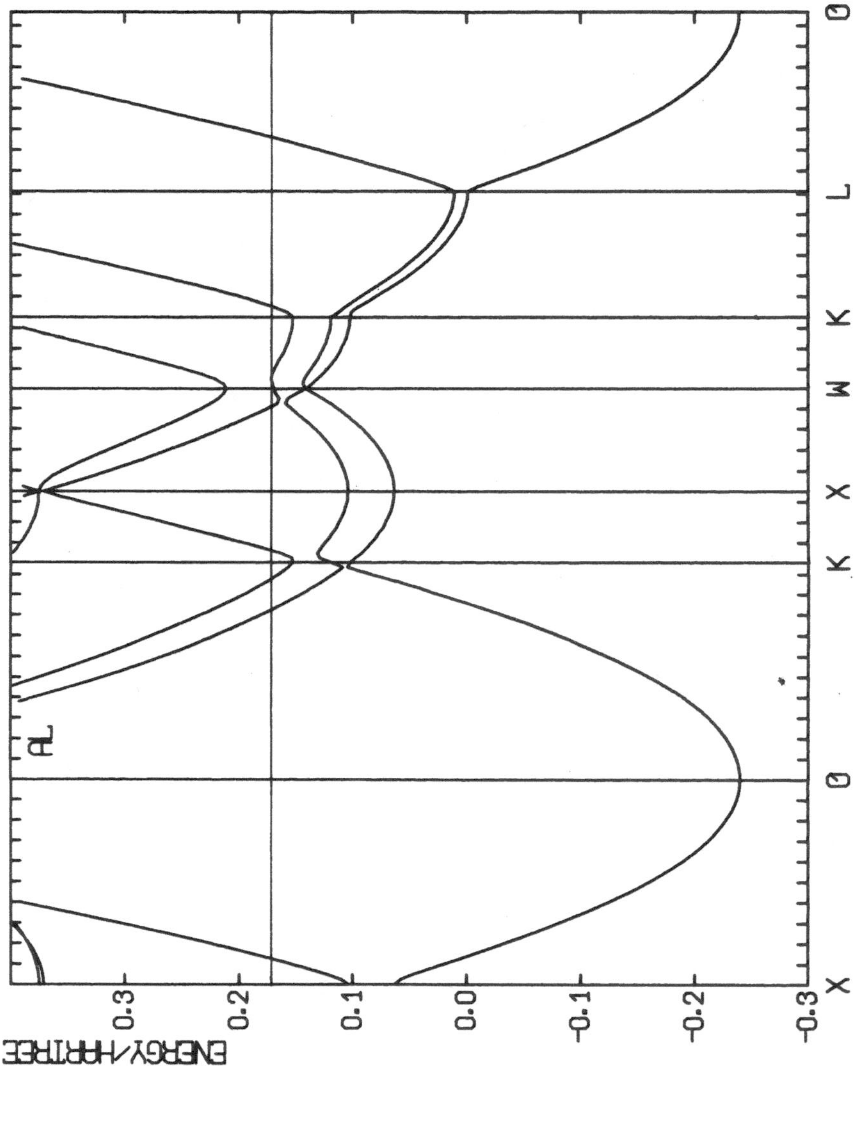
AL
ENERGY/HARTREE
0.3
0.2
0.1
0.0
-0.1
-0.2
-0.3
X
0
K
X
W
K
L
0

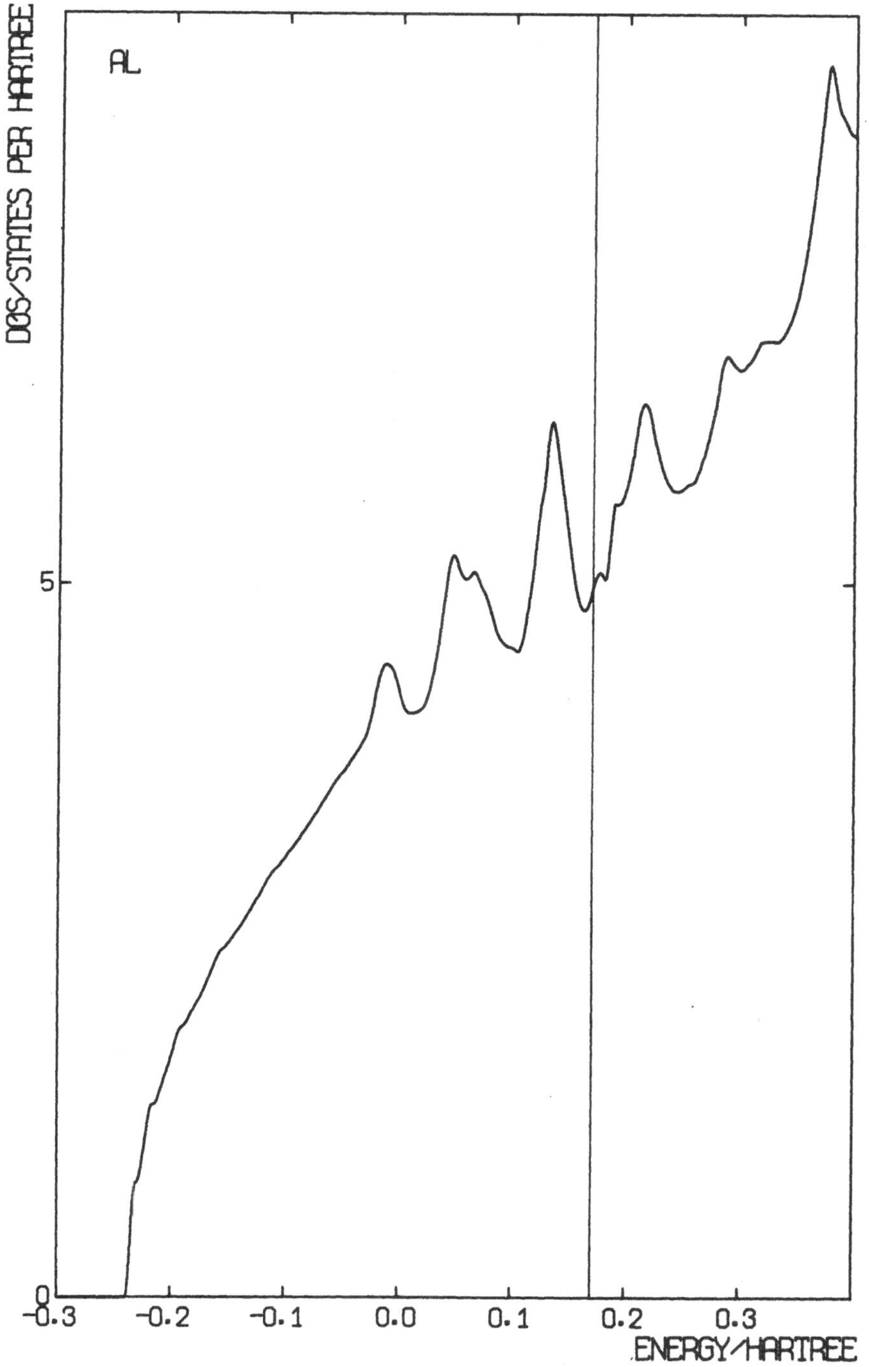

AL
DOS/STATES PER HARTREE
5
0
-0.3
-0.2
-0.1
0.0
0.1
0.2
0.3
ENERGY/HARTREE

```
POTASSIUM, B.C.C. STRUCTURE, A = 9.87 A.U.

 EXPONENTS FOR THE POTENTIAL BASIS:
 19.0000      6.8418      2.4637      0.8872
 POTENTIAL EXPANSION COEFFICIENTS:
  0.11645E+02-0.26383E+02 0.25047E+02 0.86919E+01
  0.61552E+02-0.20454E+01-0.27190E+02-0.11898E+01
  0.37231E+03-0.22195E+03 0.84510E+01 0.18042E+00

 EXPONENTS FOR THE ATOM-LIKE ORBITAL BASIS:
 19.0000      6.8418      2.4637      0.8871
 ORBITAL BASIS EXPANSION COEFFICIENTS:
   1S:
 -0.16750E+03 0.11020E+02-0.23937E+01 0.12051E+00
 -0.93172E+02-0.27707E+02 0.24800E+01-0.38203E-01
 -0.94776E+03 0.54465E+02-0.97576E+00 0.29975E-02
   2S:
 -0.48818E+02 0.40998E+00 0.28791E+01-0.92615E-01
 -0.18998E+03 0.13799E+03-0.27347E+01 0.28801E-01
 -0.54299E+03-0.25133E+02 0.92408E+00-0.22193E-02
   2P:
  0.58255E+02 0.15449E+03-0.40860E+00 0.12882E-01
  0.25726E+03-0.10424E+03 0.45770E+00-0.37235E-02
  0.99980E+03 0.12209E+03-0.15994E+00 0.26381E-03
   3S:
 -0.17967E+02-0.59930E+00 0.34610E+01-0.82862E+00
 -0.77862E+02 0.54923E+02-0.13292E+02 0.22957E+00
 -0.22768E+03 0.52523E+02 0.17473E+00-0.16142E-01
   3P:
 -0.15111E+02-0.62335E+02 0.87142E+01 0.34012E+00
 -0.66151E+02 0.16616E+02 0.27371E+00-0.77024E-01
 -0.22559E+03-0.74827E+02 0.13024E+01 0.44070E-02
   3D:
 -0.23206E+02 0.30391E+02 0.17951E+01 0.70193E-01
  0.39841E+01-0.56946E+02 0.14665E+01 0.62799E-01
 -0.14709E+04 0.10399E+03-0.94710E-01-0.62008E-02
   4S:
 -0.16166E+02 0.65140E+02-0.59100E+02 0.14598E+02
 -0.12037E+03-0.15349E+01 0.53094E+02-0.54896E+01
 -0.44591E+03 0.34702E+03-0.41441E+02 0.21352E+00
   4P:
  0.35945E+02-0.33394E+02 0.25283E+02-0.53496E+01
  0.20317E+03 0.61414E+02-0.25791E+02 0.21339E+01
  0.13296E+04-0.27018E+03 0.16971E+02-0.14696E+00

 EXPONENTS FOR THE DENSITY BASIS:
 38.0000     14.8538      5.8062      2.2696      0.8871
 SITE DENSITY EXPANSION COEFFICIENTS:
  0.60156E+05-0.76691E+04 0.26577E+04 0.49162E+02 0.36262E+00
  0.87831E+05 0.18363E+05-0.75347E+04 0.19890E+02 0.22389E+00
  0.88875E+06 0.90168E+05 0.50443E+04-0.13234E+02-0.19583E-01
 TOTAL ATOMIC SPHERE DENSITY EXPANSION COEFFICIENTS:
  0.61586E+05-0.96441E+04 0.35426E+04-0.22158E+03 0.51290E+02
  0.10068E+06 0.25078E+05-0.89200E+04 0.21275E+03-0.18742E+02
  0.11242E+07 0.49273E+05 0.74420E+04-0.12535E+03 0.18899E+01
```

SLATER-KOSTER INTEGRALS:

HAMILTONIAN INTEGRALS:

ONE-CENTRE:

(S)	(P0)	(P1-)	(P1+)
0.0229	0.0906	0.0906	0.0906

TWO-CENTRE:

(SSS)	(PSS)	(PPS)	(PPP)
-0.0117	0.0042	-0.0198	-0.0027
-0.0103	0.0095	-0.0003	-0.0040
-0.0026	0.0039	0.0066	-0.0010
-0.0007	0.0010	0.0019	-0.0002
-0.0007	0.0011	0.0021	-0.0000

OVERLAP INTEGRALS:

ONE-CENTRE:

(S)	(P0)	(P1-)	(P1+)
0.9847	0.9801	0.9801	0.9801

TWO-CENTRE:

(SSS)	(PSS)	(PPS)	(PPP)
0.1970	-0.2752	-0.3890	0.1122
0.1129	-0.1684	-0.2734	0.0511
0.0124	-0.0151	-0.0258	0.0015
0.0028	-0.0017	-0.0007	-0.0003
0.0021	-0.0013	-0.0006	-0.0005

FERMI ENERGY = 0.0344 HARTREE

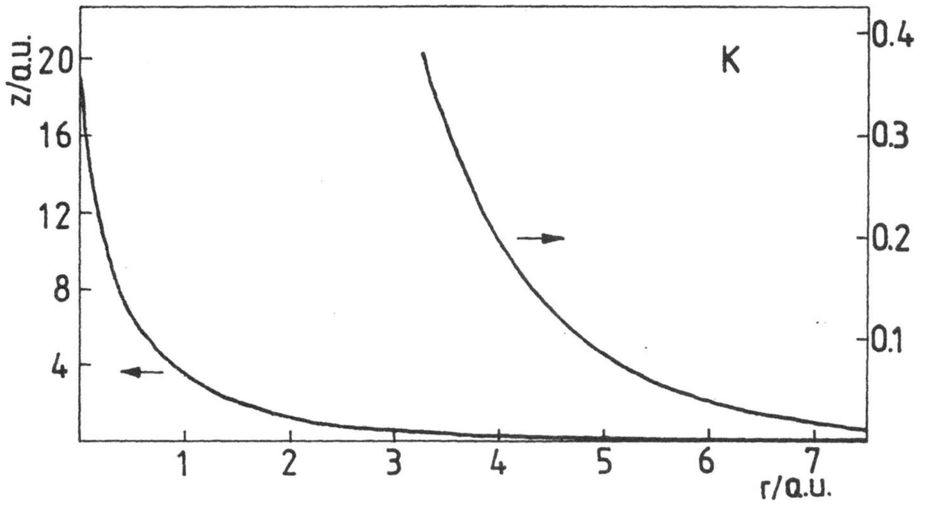

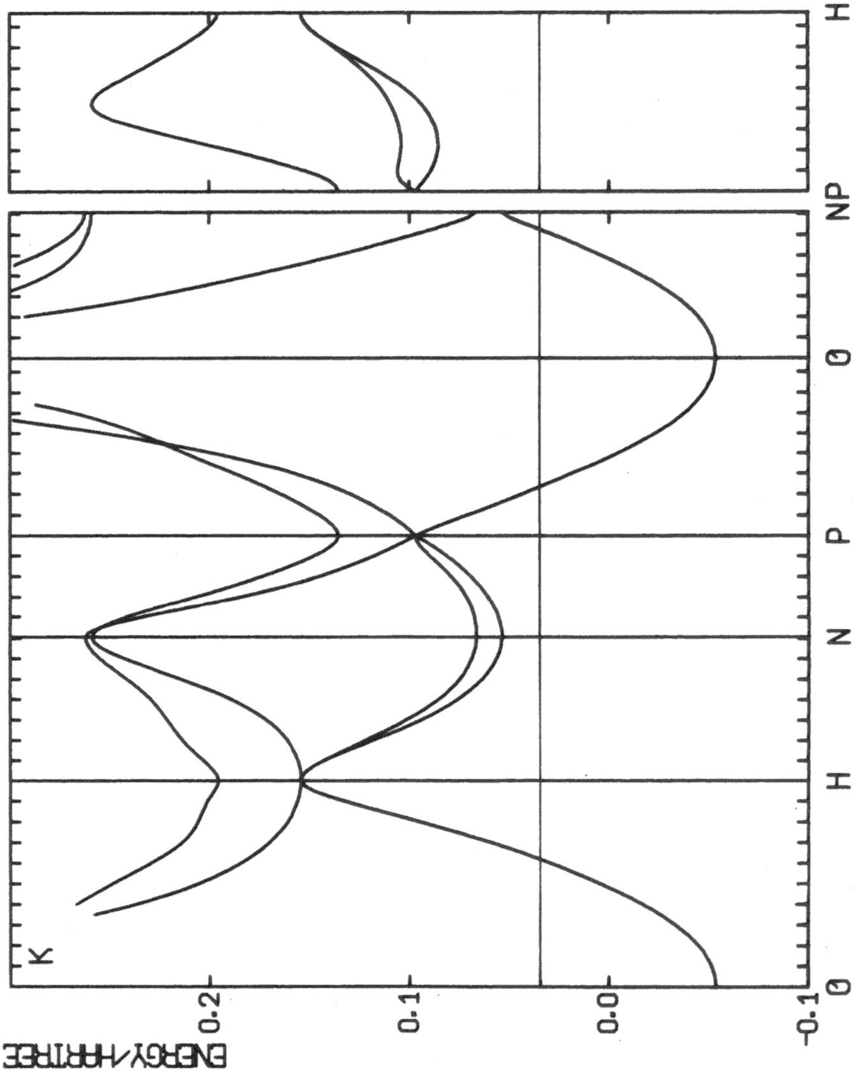
ENERGY/HARTREE
0.2
0.1
0.0
-0.10
K
0
H
N
P
0
NP
H

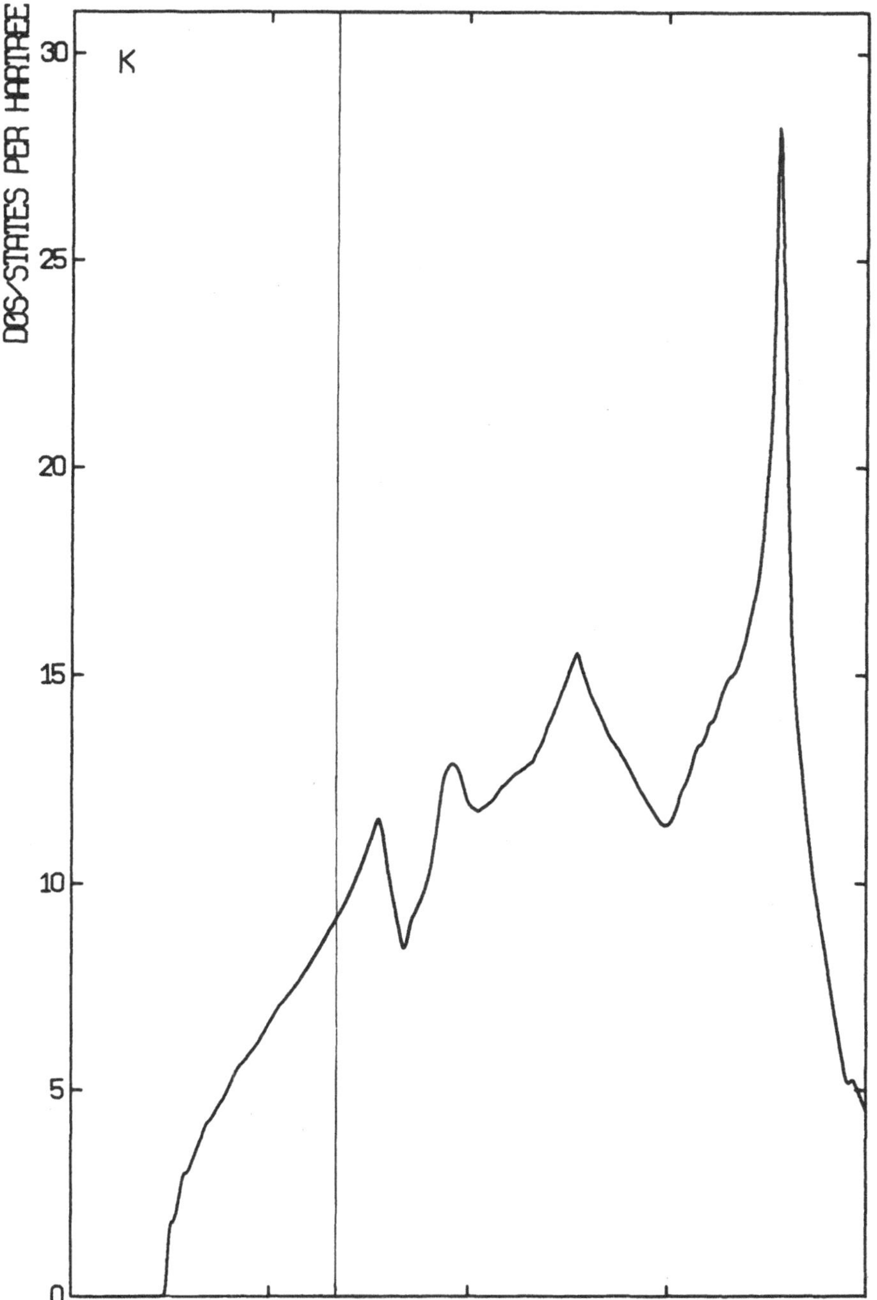
K
DOS/STATES PER HARTREE
30
25
20
15
10
5
0
-0.1
0.0
0.1
0.2
ENERGY/HARTREE

```
CALCIUM, F.C.C. STRUCTURE, A = 10.45 A.U.

 EXPONENTS FOR THE POTENTIAL BASIS:
 20.0000      7.4928       2.8071       1.0517
 POTENTIAL EXPANSION COEFFICIENTS:
  0.20232E+02-0.70058E+02 0.90203E+02-0.20377E+02
  0.12521E+03-0.54439E+02-0.57276E+02 0.14414E+02
  0.61057E+03-0.45147E+03 0.75994E+02-0.16540E+01

 EXPONENTS FOR THE ATOM-LIKE ORBITAL BASIS:
 20.0000      7.4928       2.8071       1.0517
 ORBITAL BASIS EXPANSION COEFFICIENTS:
   1S:
  0.18311E+03-0.14305E+02 0.34243E+01-0.17861E+00
  0.11338E+03 0.36203E+02-0.38858E+01 0.64183E-01
  0.10996E+04-0.81176E+02 0.17091E+01-0.57339E-02
   2S:
 -0.50707E+02-0.30846E+01 0.40484E+01-0.12102E+00
 -0.20142E+03 0.17247E+03-0.41465E+01 0.42365E-01
 -0.56356E+03-0.18827E+02 0.15112E+01-0.36912E-02
   2P:
 -0.61488E+02-0.18545E+03 0.91094E-01-0.11007E-01
 -0.28230E+03 0.11307E+03-0.23139E+00 0.36570E-02
 -0.96621E+03-0.16550E+03 0.12564E+00-0.29867E-03
   3S:
 -0.19118E+02-0.51939E+01 0.76112E+01-0.12861E+01
 -0.84709E+02 0.62122E+02-0.20654E+02 0.40052E+00
 -0.24338E+03 0.63649E+02 0.25533E+00-0.31751E-01
   3P:
 -0.14876E+02-0.80755E+02 0.11050E+02 0.53586E+00
 -0.64225E+02 0.18415E+02 0.19438E+01-0.14223E+00
 -0.15194E+03-0.12224E+03 0.19899E+01 0.95189E-02
   3D:
 -0.27144E+02 0.47454E+02 0.50630E+01 0.47504E+00
 -0.35825E+02-0.74751E+02 0.21908E+01-0.46636E-01
 -0.17352E+04 0.17505E+03-0.56907E+00-0.49694E-03
   4S:
 -0.27180E+02 0.10852E+03-0.10194E+03 0.26291E+02
 -0.19961E+03 0.19577E+02 0.94716E+02-0.11450E+02
 -0.70528E+03 0.63679E+03-0.85443E+02 0.70244E+00
   4P:
  0.35087E+02-0.29030E+02 0.32064E+02-0.83453E+01
  0.21164E+03 0.47661E+02-0.34095E+02 0.39336E+01
  0.11870E+04-0.31041E+03 0.27588E+02-0.32583E+00

 EXPONENTS FOR THE DENSITY BASIS:
 40.0000     16.1070      6.4859       2.6117       1.0517
 SITE DENSITY EXPANSION COEFFICIENTS:
  0.67940E+05-0.74359E+04 0.39699E+04 0.10881E+03-0.21139E+01
  0.88546E+05-0.58504E+04-0.12297E+05 0.26791E+02 0.20119E+01
  0.49173E+06 0.18563E+06 0.82266E+04-0.26921E+02-0.19678E+00
 TOTAL ATOMIC SPHERE DENSITY EXPANSION COEFFICIENTS:
  0.68233E+05-0.80950E+04 0.46422E+04-0.31007E+03 0.11916E+03
  0.92027E+05-0.53940E+04-0.12858E+05 0.29478E+03-0.52097E+02
  0.52785E+06 0.17006E+06 0.10657E+05-0.27972E+03 0.64052E+01
```

SLATER-KOSTER INTEGRALS:

HAMILTONIAN INTEGRALS:

ONE-CENTRE:

(S)	(PO)	(P1-)	(P1+)	(DO)	(D1-)	(D1+)	(D2-)	(D2+)
0.0212	0.1134	0.1134	0.1134	0.1448	0.1427	0.1427	0.1427	0.1448

TWO-CENTRE:

(SSS)	(PSS)	(DSS)	(PPS)	(PPP)	(DPS)	(DPP)	(DDS)	(DDP)	(DDD)
-0.0194	0.0130	0.0073	-0.0128	-0.0064	0.0006	-0.0059	-0.0051	0.0058	-0.0012
-0.0071	0.0103	0.0037	0.0146	-0.0028	0.0057	-0.0016	-0.0025	0.0009	-0.0001
-0.0018	0.0028	0.0009	0.0051	-0.0007	0.0016	-0.0004	-0.0005	0.0002	-0.0000
-0.0007	0.0011	0.0004	0.0021	-0.0000	0.0008	-0.0000	-0.0003	0.0000	0.0000

OVERLAP INTEGRALS:

ONE-CENTRE:

(S)	(PO)	(P1-)	(P1+)	(DO)	(D1-)	(D1+)	(D2-)	(D2+)
0.9820	0.9768	0.9768	0.9768	0.9987	0.9974	0.9974	0.9974	0.9987

TWO-CENTRE:

(SSS)	(PSS)	(DSS)	(PPS)	(PPP)	(DPS)	(DPP)	(DDS)	(DDP)	(DDD)
0.1737	-0.2481	-0.0884	-0.3622	0.0931	-0.1394	0.0623	0.0675	-0.0445	0.0070
0.0269	-0.0397	-0.0138	-0.0738	0.0061	-0.0280	0.0041	0.0107	-0.0026	0.0002
0.0045	-0.0038	-0.0012	-0.0052	-0.0001	-0.0021	0.0001	0.0007	-0.0001	-0.0000
0.0012	-0.0003	-0.0003	0.0007	-0.0005	-0.0002	-0.0002	0.0002	0.0001	-0.0000

FERMI ENERGY = 0.0585 HARTREE

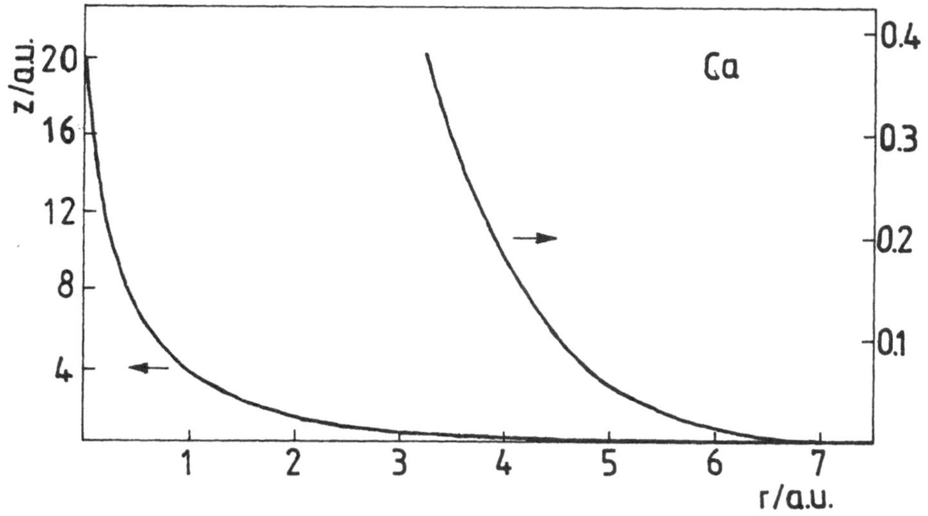

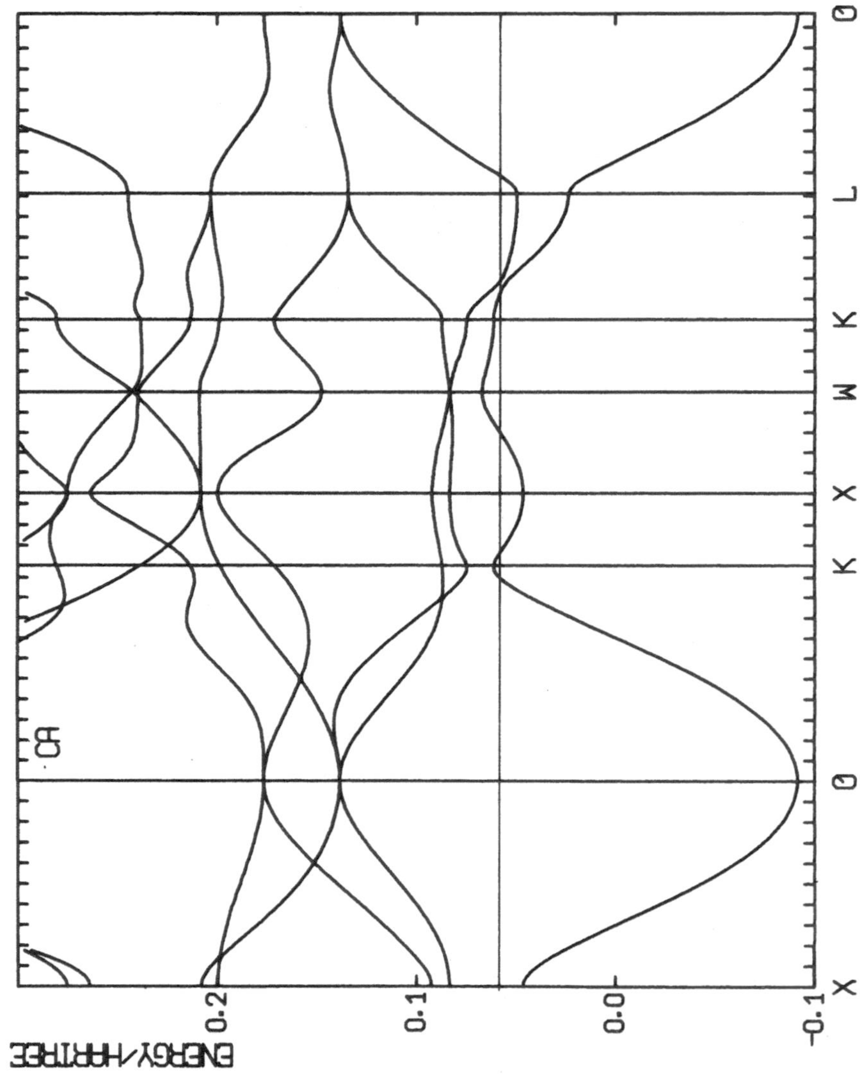
ENERGY/HARTREE
CA
0.2
0.1
0.0
-0.1
X
Ø
K
X
W
K
L
Ø

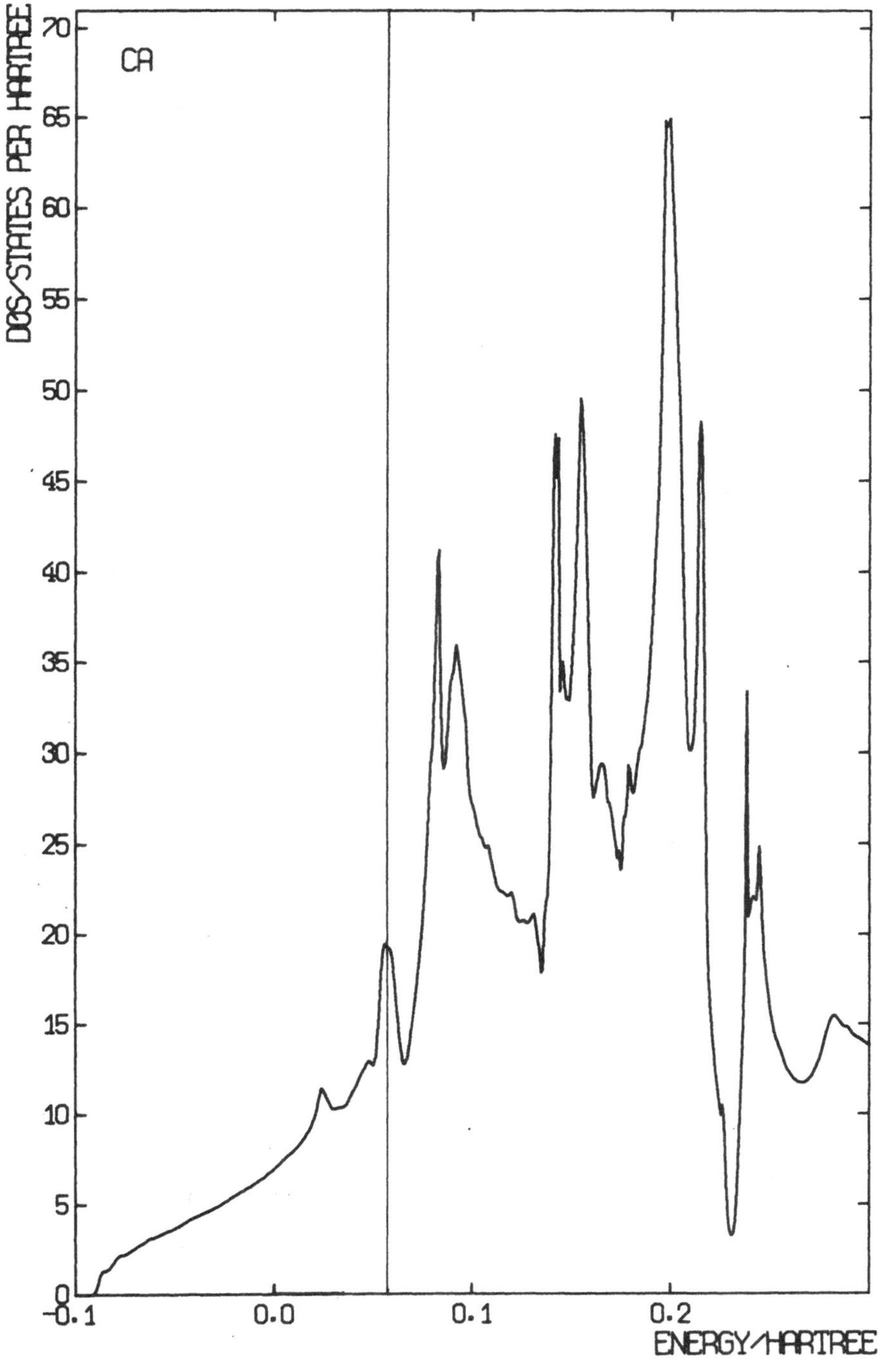
CA
DOS/STATES PER HARTREE
0
5
10
15
20
25
30
35
40
45
50
55
60
65
70
-0.1
0.0
0.1
0.2
ENERGY/HARTREE

SCANDIUM, H.C.P. STRUCTURE, A = 6.25 A.U., C = 9.96 A.U.

```
EXPONENTS FOR THE POTENTIAL BASIS:
21.0000      8.1946      3.1977      1.2478
POTENTIAL EXPANSION COEFFICIENTS:
 0.26110E+03-0.70014E+03 0.51854E+03-0.58504E+02
 0.18111E+04 0.23478E+02-0.63107E+03 0.30434E+02
 0.58038E+04-0.53285E+04 0.45652E+03-0.25837E+01

EXPONENTS FOR THE ATOM-LIKE ORBITAL BASIS:
21.0000      8.1946      3.1977      1.2478
ORBITAL BASIS EXPANSION COEFFICIENTS:
  1S:
 0.19554E+03-0.13997E+02 0.39554E+01-0.21942E+00
 0.11998E+03 0.34137E+02-0.49131E+01 0.89326E-01
 0.10299E+04-0.93010E+02 0.24327E+01-0.90814E-02
  2S:
-0.49709E+02-0.12989E+02 0.89455E+01-0.38039E+00
-0.19699E+03 0.21450E+03-0.10336E+02 0.15421E+00
-0.53213E+03-0.46933E+02 0.46332E+01-0.15624E-01
  2P:
-0.45864E+02-0.24037E+03 0.19367E+01-0.69815E-01
-0.16804E+03 0.17773E+03-0.25952E+01 0.26111E-01
 0.75399E+02-0.33252E+03 0.11436E+01-0.24144E-02
  3S:
-0.22335E+02-0.30042E+01 0.72672E+01-0.18390E+01
-0.10701E+03 0.69836E+02-0.20358E+02 0.58804E+00
-0.29722E+03 0.13888E+03-0.50179E+01-0.47947E-01
  3P:
-0.18133E+02-0.96391E+02 0.13309E+02 0.95851E+00
-0.80911E+02-0.44726E+01 0.38144E+01-0.27712E+00
-0.22075E+03-0.16004E+03 0.39198E+01 0.20106E-01
  3D:
-0.47812E+02 0.80841E+02 0.69503E+01 0.74074E+00
-0.19282E+03-0.15440E+03 0.55436E+01-0.68030E-01
-0.27492E+04 0.36834E+03-0.12631E+01-0.29158E-02
  4S:
-0.55879E+02 0.21360E+03-0.20250E+03 0.51627E+02
-0.40040E+03 0.84151E+02 0.20013E+03-0.24886E+02
-0.13478E+04 0.13920E+04-0.20446E+03 0.19683E+01
  4P:
 0.28901E+02-0.20580E+02 0.45382E+02-0.15030E+02
 0.17928E+03 0.42138E+01-0.46305E+02 0.80646E+01
 0.76527E+03-0.32455E+03 0.50590E+02-0.78794E+00

EXPONENTS FOR THE DENSITY BASIS:
42.0000     17.4371      7.2393      3.0055      1.2478
SITE DENSITY EXPANSION COEFFICIENTS:
 0.81730E+05-0.13490E+05 0.72759E+04-0.61604E+02 0.16092E+02
 0.13382E+06-0.11800E+05-0.22118E+05 0.32132E+03-0.37722E+01
 0.12261E+07 0.18724E+06 0.16833E+05-0.18511E+03 0.35523E+00
TOTAL ATOMIC SPHERE DENSITY EXPANSION COEFFICIENTS:
 0.92879E+05-0.31160E+05 0.17037E+05-0.33682E+04 0.62019E+03
 0.26071E+06 0.39535E+05-0.38537E+05 0.32621E+04-0.29158E+03
 0.29555E+07-0.27285E+06 0.54381E+05-0.23153E+04 0.37404E+02
```

SLATER-KOSTER INTEGRALS:

HAMILTONIAN INTEGRALS:

ONE-CENTRE:

(S)	(P0)	(P1-)	(P1+)	(D0)	(D1-)	(D1+)	(D2-)	(D2+)
-0.0208	0.0984	0.1008	0.1008	0.0865	0.0883	0.0883	0.0883	0.0883

TWO-CENTRE:

(SSS)	(PSS)	(DSS)	(PPS)	(PPP)	(DPS)	(DPP)	(DDS)	(DDP)	(DDD)
-0.0338	0.0279	0.0152	-0.0021	-0.0119	0.0114	-0.0112	-0.0132	0.0105	-0.0019
-0.0342	0.0267	0.0154	-0.0057	-0.0119	0.0108	-0.0117	-0.0138	0.0113	-0.0021
-0.0124	0.0180	0.0064	0.0256	-0.0043	0.0100	-0.0024	-0.0042	0.0014	-0.0002
-0.0057	0.0084	0.0026	0.0139	-0.0022	0.0044	-0.0011	-0.0016	0.0006	-0.0001
-0.0028	0.0042	0.0013	0.0072	-0.0011	0.0021	-0.0006	-0.0007	0.0003	-0.0000
-0.0030	0.0045	0.0013	0.0077	-0.0012	0.0022	-0.0006	-0.0007	0.0003	-0.0000
-0.0015	0.0023	0.0008	0.0041	-0.0003	0.0013	-0.0002	-0.0004	0.0001	-0.0000
-0.0011	0.0017	0.0007	0.0031	-0.0000	0.0012	-0.0000	-0.0004	0.0000	0.0000

OVERLAP INTEGRALS:

ONE-CENTRE:

(S)	(P0)	(P1-)	(P1+)	(D0)	(D1-)	(D1+)	(D2-)	(D2+)
0.9680	0.9597	0.9637	0.9637	0.9962	0.9971	0.9971	0.9969	0.9969

TWO-CENTRE:

(SSS)	(PSS)	(DSS)	(PPS)	(PPP)	(DPS)	(DPP)	(DDS)	(DDP)	(DDD)
0.1793	-0.2509	-0.0879	-0.3611	0.0917	-0.1353	0.0606	0.0620	-0.0412	0.0066
0.1890	-0.2619	-0.0915	-0.3709	0.0992	-0.1385	0.0654	0.0648	-0.0447	0.0074
0.0287	-0.0421	-0.0151	-0.0770	0.0061	-0.0292	0.0043	0.0109	-0.0028	0.0003
0.0094	-0.0106	-0.0038	-0.0186	0.0010	-0.0074	0.0009	0.0028	-0.0006	0.0001
0.0042	-0.0032	-0.0013	-0.0041	-0.0001	-0.0021	0.0002	0.0008	-0.0002	0.0000
0.0045	-0.0035	-0.0014	-0.0047	-0.0000	-0.0023	0.0002	0.0009	-0.0002	0.0000
0.0019	-0.0007	-0.0005	0.0004	-0.0005	-0.0005	-0.0001	0.0003	-0.0000	-0.0000
0.0012	-0.0005	-0.0005	0.0003	-0.0005	-0.0005	-0.0002	0.0004	0.0001	-0.0000

FERMI ENERGY = 0.0258 HARTREE

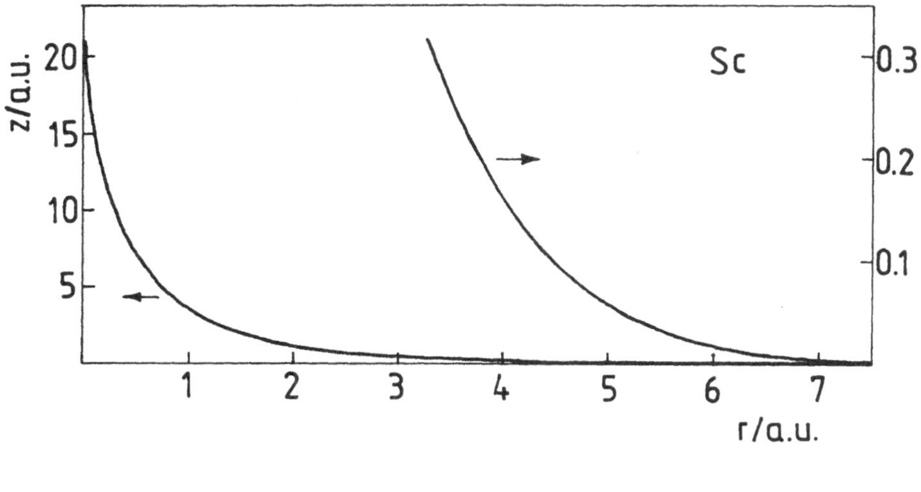

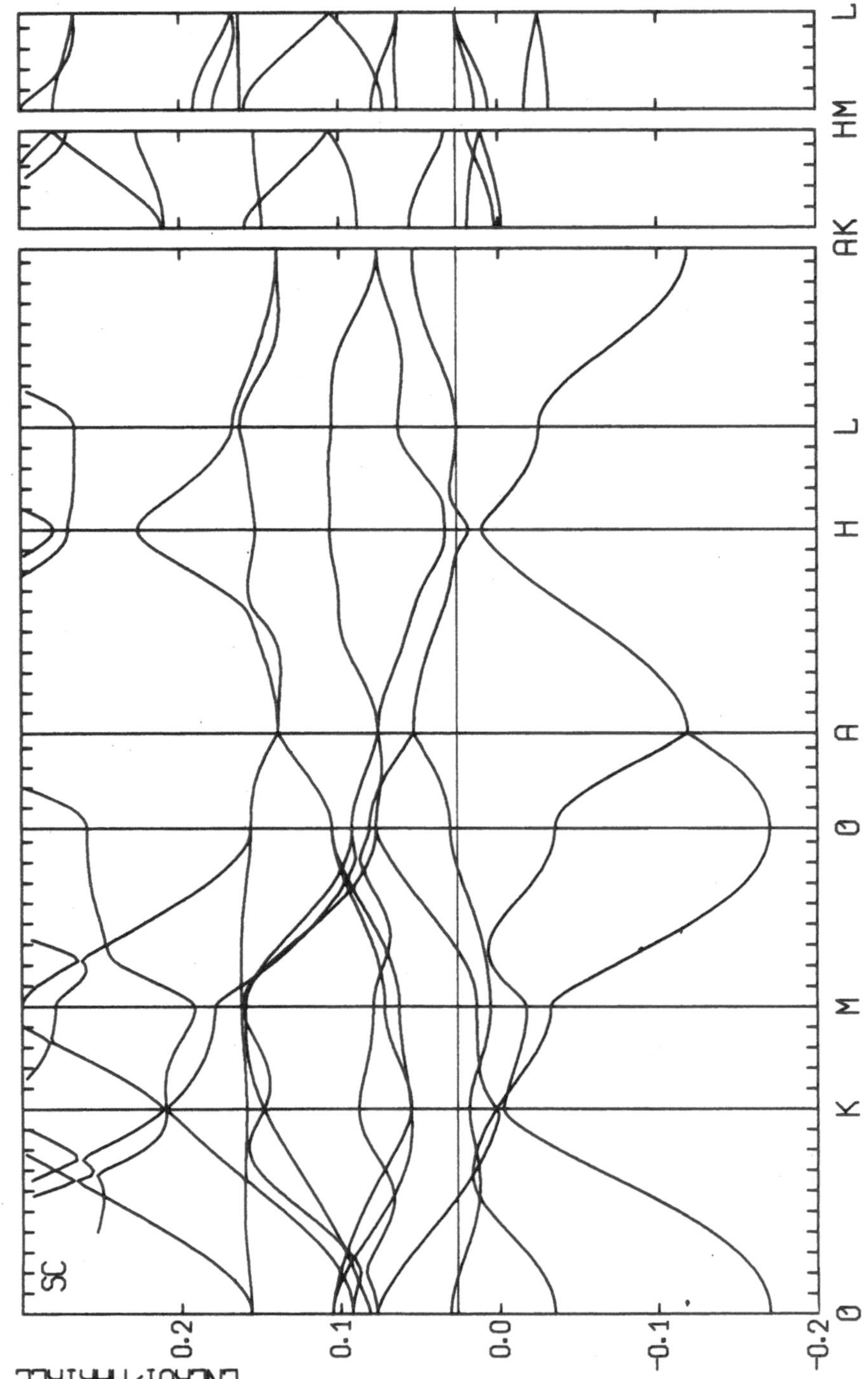
ENERGY/HARTREE
0.2
0.1
0.0
-0.1
-0.2
SC
0
K
M
0
A
H
L
L
AK
HM
L

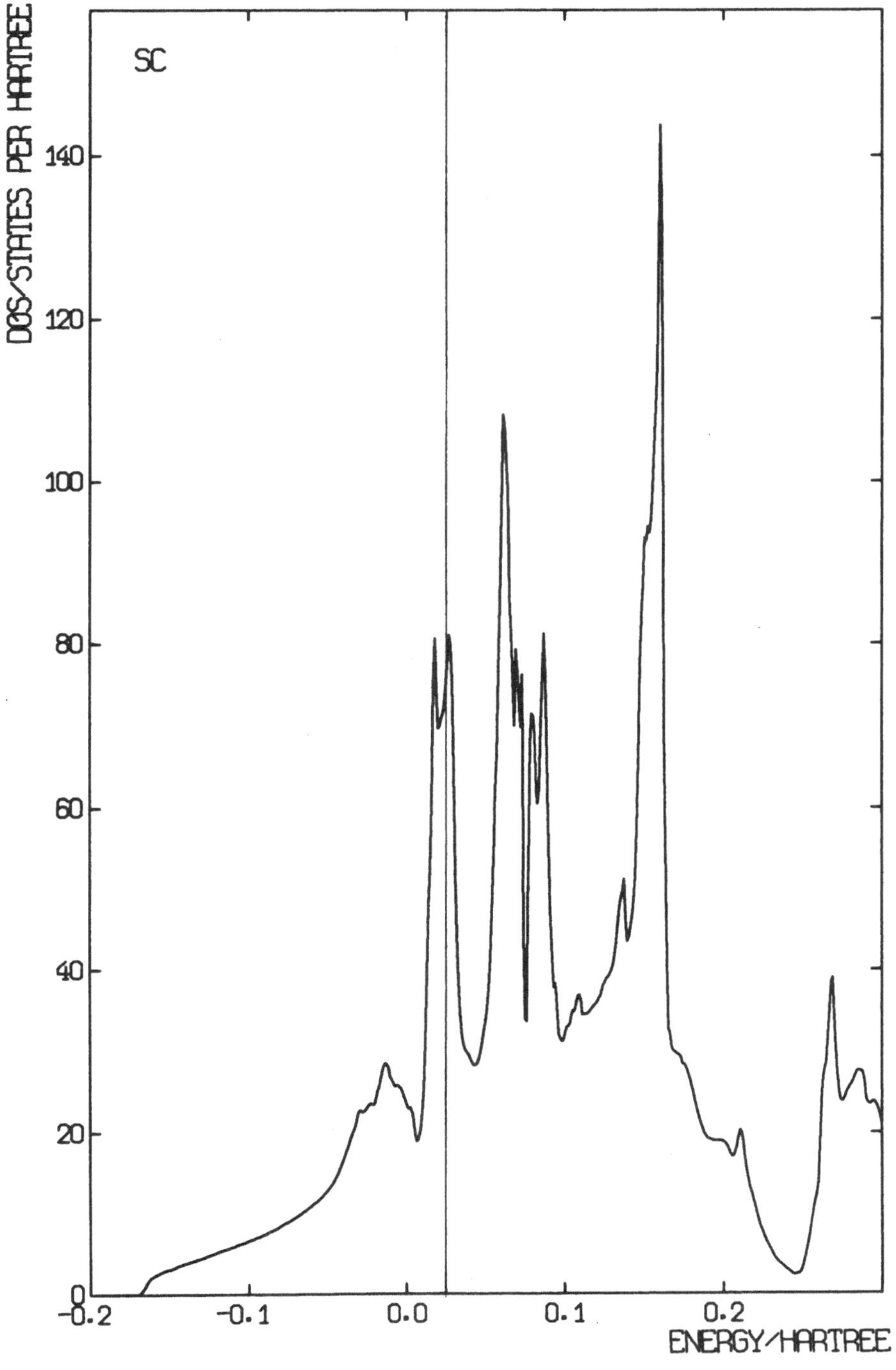
DOS/STATES PER HARTREE
SC
140
120
100
80
60
40
20
0
-0.2
-0.1
0.0
0.1
0.2
ENERGY/HARTREE

```
TITANIUM, H.C.P. STRUCTURE, A = 5.57 A.U., C = 8.84 A.U.

 EXPONENTS FOR THE POTENTIAL BASIS:
 22.0000       8.7774        3.5019        1.3972
 POTENTIAL EXPANSION COEFFICIENTS:
 -0.76355E+03 0.24006E+04-0.20025E+04 0.38750E+03
 -0.53803E+04 0.92143E+03 0.23897E+04-0.19580E+03
 -0.15937E+05 0.19097E+05-0.21808E+04 0.27370E+02

 EXPONENTS FOR THE ATOM-LIKE ORBITAL BASIS:
 22.0000       8.7774        3.5019        1.3972
 ORBITAL BASIS EXPANSION COEFFICIENTS:
   1S:
  0.22896E+03-0.36806E+02 0.92551E+01-0.48661E+00
  0.21914E+03 0.97583E+02-0.12207E+02 0.21479E+00
  0.22321E+04-0.25438E+03 0.63725E+01-0.23735E-01
   2S:
 -0.61105E+02 0.24761E+01-0.95126E+00 0.43642E+00
 -0.26018E+03 0.23748E+03 0.27513E+01-0.20264E+00
 -0.74885E+03 0.13885E+03-0.32285E+01 0.23485E-01
   2P:
  0.12098E+03 0.19476E+03 0.10315E+02-0.24331E+00
  0.72503E+03 0.75181E+02-0.12587E+02 0.98614E-01
  0.38645E+04-0.92074E+02 0.51998E+01-0.99143E-02
   3S:
 -0.19084E+02-0.27947E+02 0.29867E+02-0.47888E+01
 -0.82892E+02 0.51112E+02-0.52175E+02 0.18676E+01
 -0.27608E+03-0.15370E+02 0.12427E+02-0.18728E+00
   3P:
 -0.38985E+01 0.13968E+03-0.18615E+02-0.17045E+01
 -0.64740E+02-0.84632E+02-0.27404E+01 0.56374E+00
 -0.56566E+03 0.42810E+03-0.49284E+01-0.47245E-01
   3D:
 -0.16484E+02 0.47147E+02 0.20880E+02 0.10102E+01
  0.15859E+03-0.13126E+02-0.55966E+01-0.88135E-01
 -0.20787E+04 0.90562E+02 0.25849E+01-0.72526E-02
   4S:
  0.79327E+02-0.30430E+03 0.29154E+03-0.74373E+02
  0.57740E+03-0.15872E+03-0.30344E+03 0.38689E+02
  0.19150E+04-0.21814E+04 0.33883E+03-0.34575E+01
   4P:
  0.18788E+02-0.58623E+01 0.55179E+02-0.21525E+02
  0.11664E+03-0.54920E+02-0.53451E+02 0.12656E+02
  0.36472E+03-0.27390E+03 0.74732E+02-0.13734E+01

 EXPONENTS FOR THE DENSITY BASIS:
 44.0000      18.5738        7.8406        3.3098        1.3972
 SITE DENSITY EXPANSION COEFFICIENTS:
  0.12557E+06-0.54428E+05 0.20074E+05-0.85441E+03 0.58943E+02
  0.45661E+06 0.16179E+06-0.58341E+05 0.13066E+04-0.21451E+02
  0.69564E+07-0.82268E+06 0.55933E+05-0.64395E+03 0.22992E+01
 TOTAL ATOMIC SPHERE DENSITY EXPANSION COEFFICIENTS:
  0.16090E+06-0.11185E+06 0.52349E+05-0.11261E+05 0.17479E+04
  0.88501E+06 0.32713E+06-0.11707E+06 0.11697E+05-0.88014E+03
  0.12563E+08-0.24719E+07 0.19661E+06-0.82693E+04 0.11926E+03
```

SLATER-KOSTER INTEGRALS:

HAMILTONIAN INTEGRALS:

ONE-CENTRE:

(S)	(P0)	(P1-)	(P1+)	(D0)	(D1-)	(D1+)	(D2-)	(D2+)
-0.0622	0.0804	0.0839	0.0839	0.0194	0.0208	0.0208	0.0208	0.0208

TWO-CENTRE:

(SSS)	(PSS)	(DSS)	(PPS)	(PPP)	(DPS)	(DPP)	(DDS)	(DDP)	(DDD)
-0.0404	-0.0353	-0.0194	-0.0053	-0.0136	0.0167	-0.0142	-0.0186	0.0135	-0.0025
-0.0410	-0.0348	-0.0192	-0.0112	-0.0136	0.0156	-0.0149	-0.0197	0.0147	-0.0028
-0.0166	-0.0241	-0.0084	0.0338	-0.0057	0.0133	-0.0030	-0.0055	0.0017	-0.0003
-0.0079	-0.0115	-0.0035	0.0186	-0.0032	0.0057	-0.0015	-0.0020	0.0007	-0.0001
-0.0039	-0.0057	-0.0016	0.0095	-0.0015	0.0026	-0.0007	-0.0008	0.0004	-0.000[illegible]
-0.0042	-0.0061	-0.0017	0.0102	-0.0016	0.0028	-0.0008	-0.0008	0.0004	-0.0001
-0.0022	-0.0033	-0.0010	0.0058	-0.0005	0.0017	-0.0003	-0.0005	0.0002	-0.0000
-0.0015	-0.0024	-0.0009	0.0043	-0.0000	0.0015	-0.0000	-0.0006	0.0000	0.0000

OVERLAP INTEGRALS:

ONE-CENTRE:

(S)	(P0)	(P1-)	(P1+)	(D0)	(D1-)	(D1+)	(D2-)	(D2+)
0.9549	0.9467	0.9525	0.9525	0.9954	0.9964	0.9964	0.9963	0.9963

TWO-CENTRE:

(SSS)	(PSS)	(DSS)	(PPS)	(PPP)	(DPS)	(DPP)	(DDS)	(DDP)	(DDD)
0.1856	0.2551	0.0880	-0.3618	0.0931	-0.1346	0.0603	0.0614	-0.0403	0.0064
0.1969	0.2673	0.0919	-0.3721	0.1019	-0.1381	0.0660	0.0646	-0.0444	0.0073
0.0307	0.0446	0.0154	-0.0801	0.0066	-0.0295	0.0043	0.0107	-0.0027	0.0003
0.0103	0.0117	0.0039	-0.0202	0.0015	-0.0075	0.0011	0.0027	-0.0007	0.0001
0.0045	0.0035	0.0013	-0.0045	0.0001	-0.0020	0.0003	0.0007	-0.0002	0.0000
0.0049	0.0039	0.0014	-0.0052	0.0002	-0.0022	0.0004	0.0008	-0.0003	0.0000
0.0023	0.0011	0.0006	-0.0002	-0.0005	-0.0007	-0.0001	0.0003	-0.0000	-0.0000
0.0015	0.0008	0.0006	-0.0003	-0.0006	-0.0008	-0.0002	0.0004	0.0001	-0.0000

FERMI ENERGY = -0.0061 HARTREE

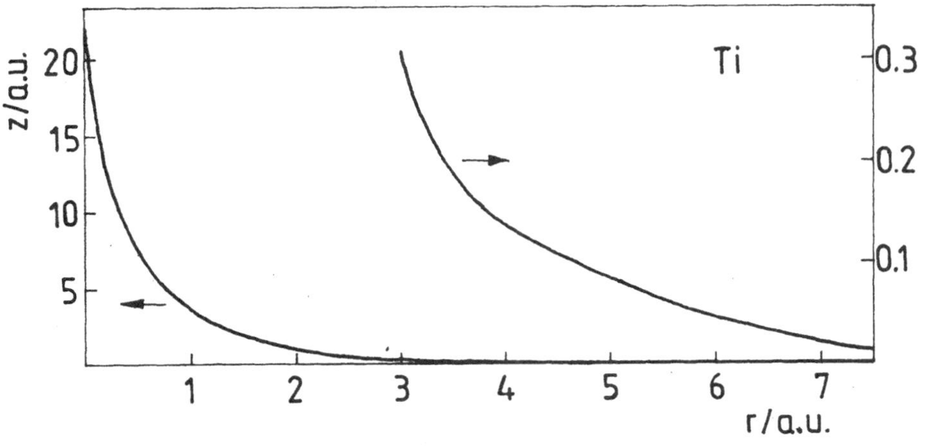

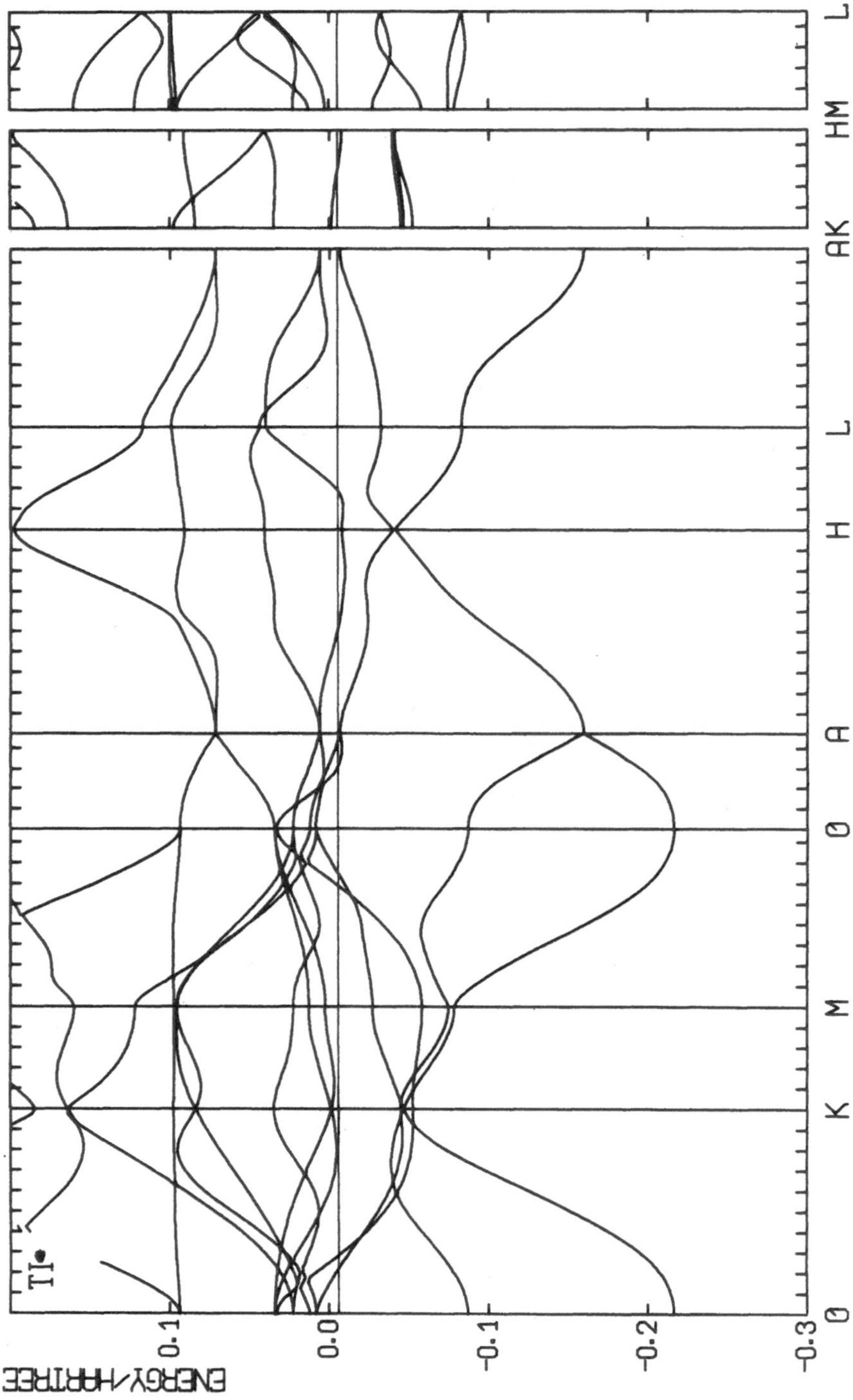
ENERGY/HARTREE
0.1
0.0
-0.1
-0.2
-0.3
Tl
Θ
K
M
Θ
A
H
L
AK
HM
L

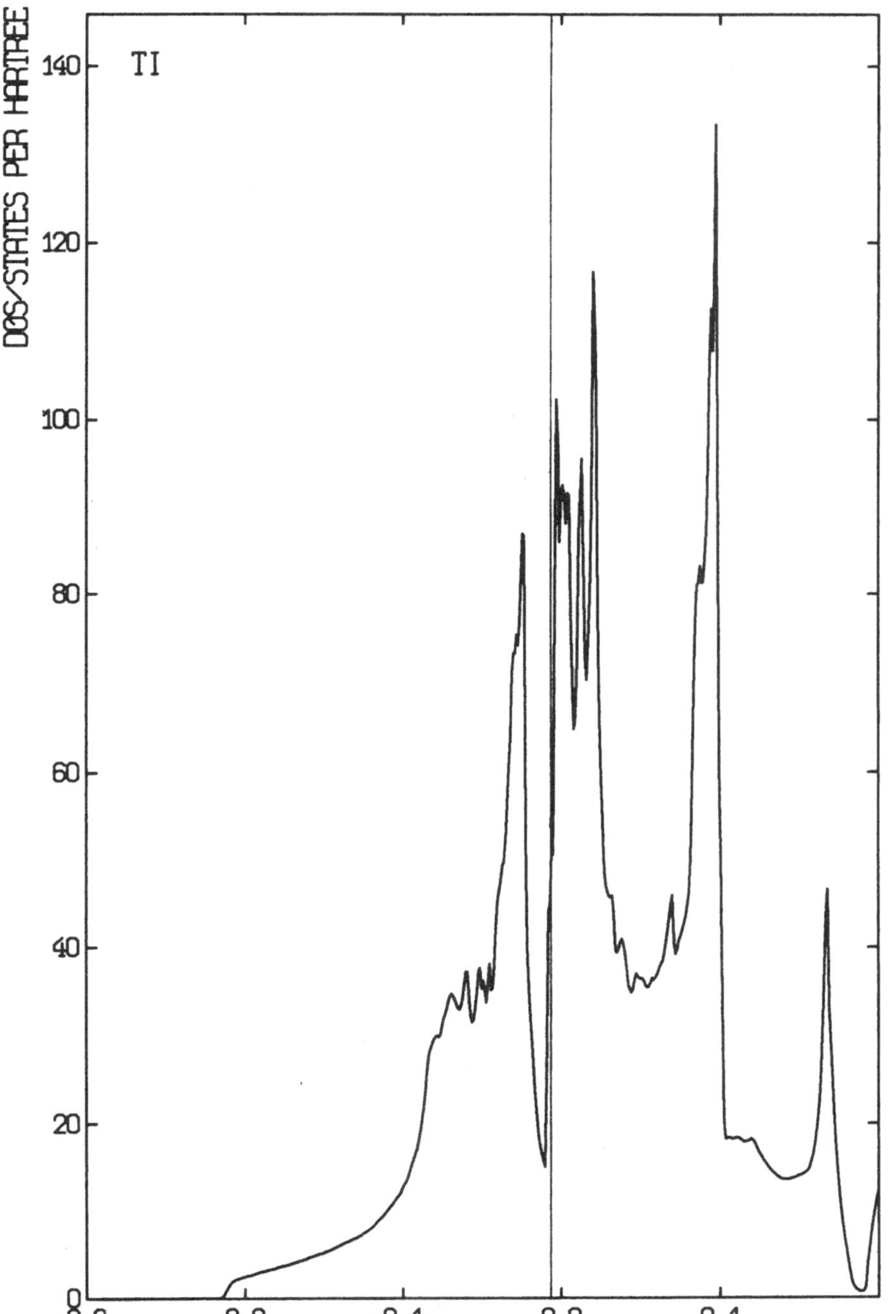
TI
DOS/STATES PER HARTREE
140
120
100
80
60
40
20
0
-0.3
-0.2
-0.1
0.0
0.1
ENERGY/HARTREE

```
VANADIUM, B.C.C. STRUCTURE, A = 5.71 A.U.
 EXPONENTS FOR THE POTENTIAL BASIS:
 23.0000      9.2811      3.7452      1.5113
 POTENTIAL EXPANSION COEFFICIENTS:
 -0.46271E+03 0.15699E+04-0.12976E+04 0.21342E+03
 -0.33160E+04 0.94607E+03 0.17742E+04-0.99779E+02
 -0.94951E+04 0.13650E+05-0.15723E+04 0.12461E+02

 EXPONENTS FOR THE ATOM-LIKE ORBITAL BASIS:
 23.0000      9.2811      3.7452      1.5113
 ORBITAL BASIS EXPANSION COEFFICIENTS:
   1S:
  0.23935E+03-0.33765E+02 0.91795E+01-0.49811E+00
  0.22303E+03 0.89941E+02-0.12815E+02 0.23484E+00
  0.21100E+04-0.26220E+03 0.71639E+01-0.27752E-01
   2S:
 -0.57261E+02-0.11365E+02 0.50336E+01 0.73247E-01
 -0.23745E+03 0.29855E+03-0.54165E+01-0.42602E-01
 -0.63909E+03 0.44001E+02 0.12242E+01 0.59647E-02
   2P:
  0.10283E+03 0.25900E+03 0.75385E+01-0.17217E+00
  0.61082E+03-0.25003E+02-0.95572E+01 0.74474E-01
  0.24696E+04 0.97928E+02 0.41305E+01-0.80017E-02
   3S:
  0.21153E+02 0.27115E+02-0.29595E+02 0.51556E+01
  0.99671E+02-0.66019E+02 0.55290E+02-0.21087E+01
  0.31566E+03-0.34822E+02-0.11255E+02 0.22084E+00
   3P:
 -0.54900E+01 0.16439E+03-0.23479E+02-0.19601E+01
 -0.84135E+02-0.88051E+02-0.16502E+01 0.68022E+00
 -0.83283E+03 0.54631E+03-0.77800E+01-0.59643E-01
   3D:
 -0.44967E+02 0.90386E+02 0.23345E+02 0.13879E+01
 -0.12274E+02-0.11500E+03-0.23188E+01-0.18358E+00
 -0.40120E+04 0.34158E+03 0.13572E+01-0.36824E-02
   4S:
  0.84613E+02-0.34779E+03 0.35561E+03-0.10144E+03
  0.63385E+03-0.23917E+03-0.35879E+03 0.58440E+02
  0.21313E+04-0.26554E+04 0.46972E+03-0.64725E+01
   4P:
  0.82070E+01 0.93969E+01 0.63558E+02-0.27485E+02
  0.42587E+02-0.11637E+03-0.59877E+02 0.17240E+02
 -0.20986E+03-0.20706E+03 0.10177E+03-0.20131E+01

 EXPONENTS FOR THE DENSITY BASIS:
 46.0000     19.5841      8.3378      3.5497      1.5113
 SITE DENSITY EXPANSION COEFFICIENTS:
  0.21330E+06-0.14862E+06 0.44934E+05-0.23272E+04 0.36844E+02
  0.13836E+07 0.65043E+06-0.13310E+06 0.32560E+04-0.44957E+01
  0.22623E+08-0.34935E+07 0.14688E+06-0.13013E+04-0.24764E+00
 TOTAL ATOMIC SPHERE DENSITY EXPANSION COEFFICIENTS:
  0.25053E+06-0.21169E+06 0.83342E+05-0.15905E+05 0.24263E+04
  0.18605E+07 0.82021E+06-0.20225E+06 0.17308E+05-0.13184E+04
  0.28766E+08-0.54757E+07 0.33566E+06-0.12841E+05 0.19407E+03
```

SLATER-KOSTER INTEGRALS:

HAMILTONIAN INTEGRALS:

ONE-CENTRE:

(S)	(P0)	(P1-)	(P1+)	(D0)	(D1-)	(D1+)	(D2-)	(D2+)
0.1278	0.2607	0.2607	0.2607	0.1715	0.1667	0.1667	0.1667	0.1715

TWO-CENTRE:

(SSS)	(PSS)	(DSS)	(PPS)	(PPP)	(DPS)	(DPP)	(DDS)	(DDP)	(DDD)
-0.0194	0.0040	-0.0119	-0.0724	-0.0040	0.0011	-0.0121	-0.0167	0.0140	-0.0028
-0.0254	-0.0205	-0.0137	-0.0110	-0.0106	0.0115	-0.0088	-0.0110	0.0071	-0.0011
-0.0065	-0.0115	-0.0031	0.0200	-0.0038	0.0058	-0.0016	-0.0018	0.0007	-0.0001
-0.0015	-0.0029	-0.0007	0.0061	-0.0009	0.0014	-0.0004	-0.0003	0.0002	-0.0000
-0.0019	-0.0039	-0.0013	0.0084	-0.0002	0.0027	-0.0001	-0.0009	0.0000	0.0000

OVERLAP INTEGRALS:

ONE-CENTRE:

(S)	(P0)	(P1-)	(P1+)	(D0)	(D1-)	(D1+)	(D2-)	(D2+)
0.9587	0.9452	0.9452	0.9452	0.9991	0.9946	0.9946	0.9946	0.9991

TWO-CENTRE:

(SSS)	(PSS)	(DSS)	(PPS)	(PPP)	(DPS)	(DPP)	(DDS)	(DDP)	(DDD)
0.1896	0.2796	0.0929	-0.3858	0.1143	-0.1348	0.0692	0.0621	-0.0438	0.0074
0.1043	0.1729	0.0584	-0.2808	0.0508	-0.0993	0.0306	0.0381	-0.0183	0.0023
0.0059	0.0102	0.0033	-0.0248	0.0017	-0.0083	0.0011	0.0026	-0.0006	0.0001
0.0004	-0.0005	0.0001	0.0003	-0.0002	-0.0003	0.0001	0.0001	-0.0001	0.0000
0.0011	0.0010	0.0008	-0.0018	-0.0007	-0.0015	-0.0002	0.0007	0.0001	-0.0000

FERMI ENERGY = 0.1378 HARTREE

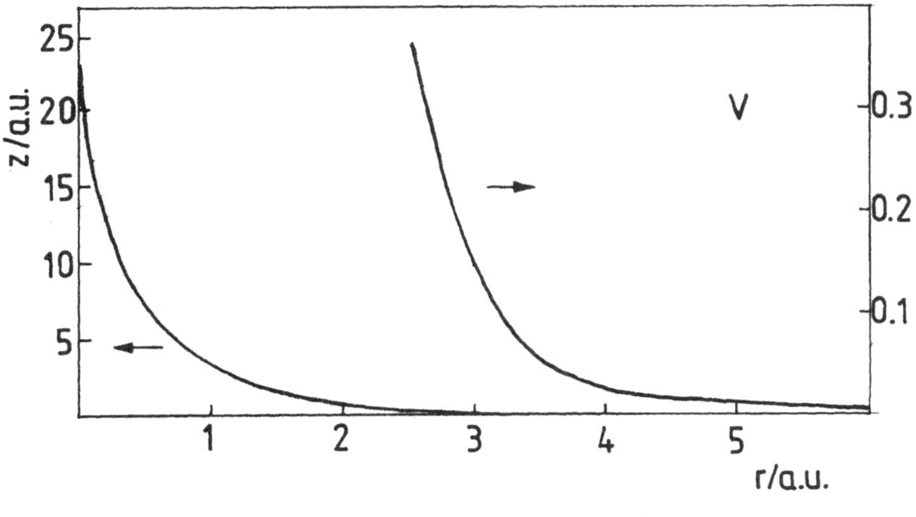

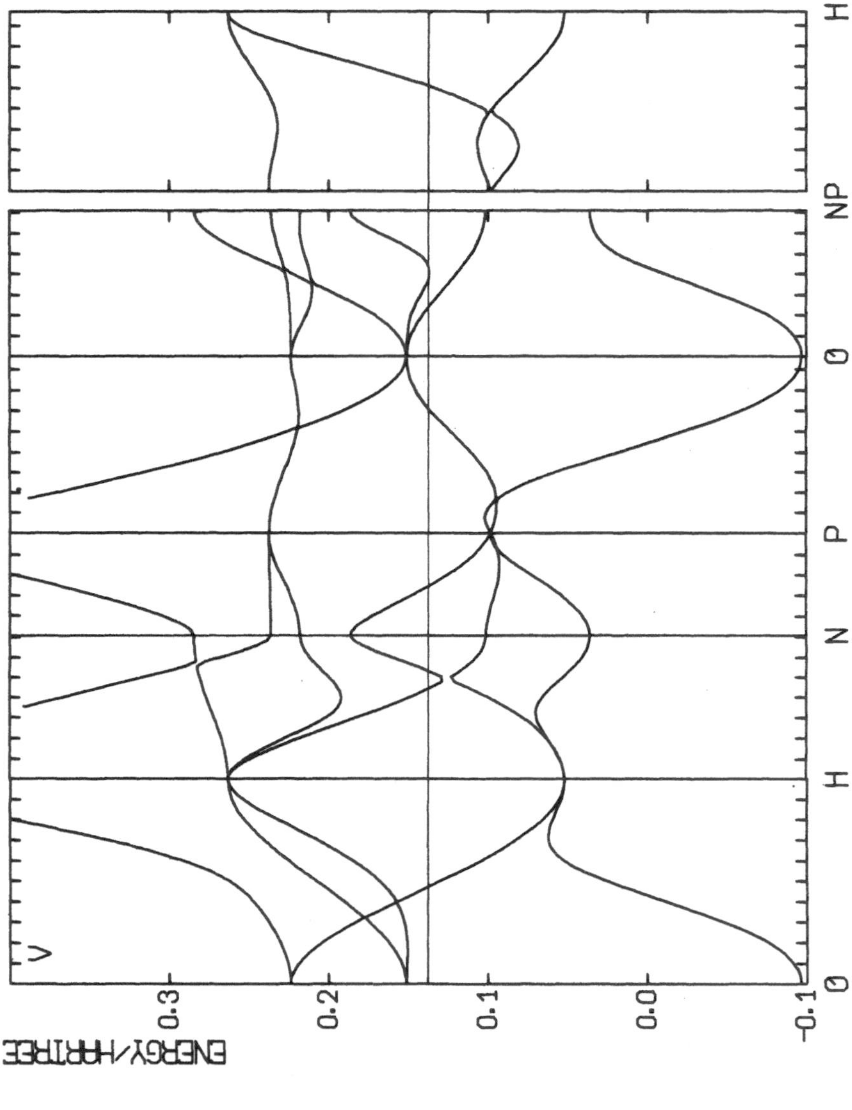
V
ENERGY/HARTREE
-0.1
0.0
0.1
0.2
0.3
0
H
N
P
0
NP
H

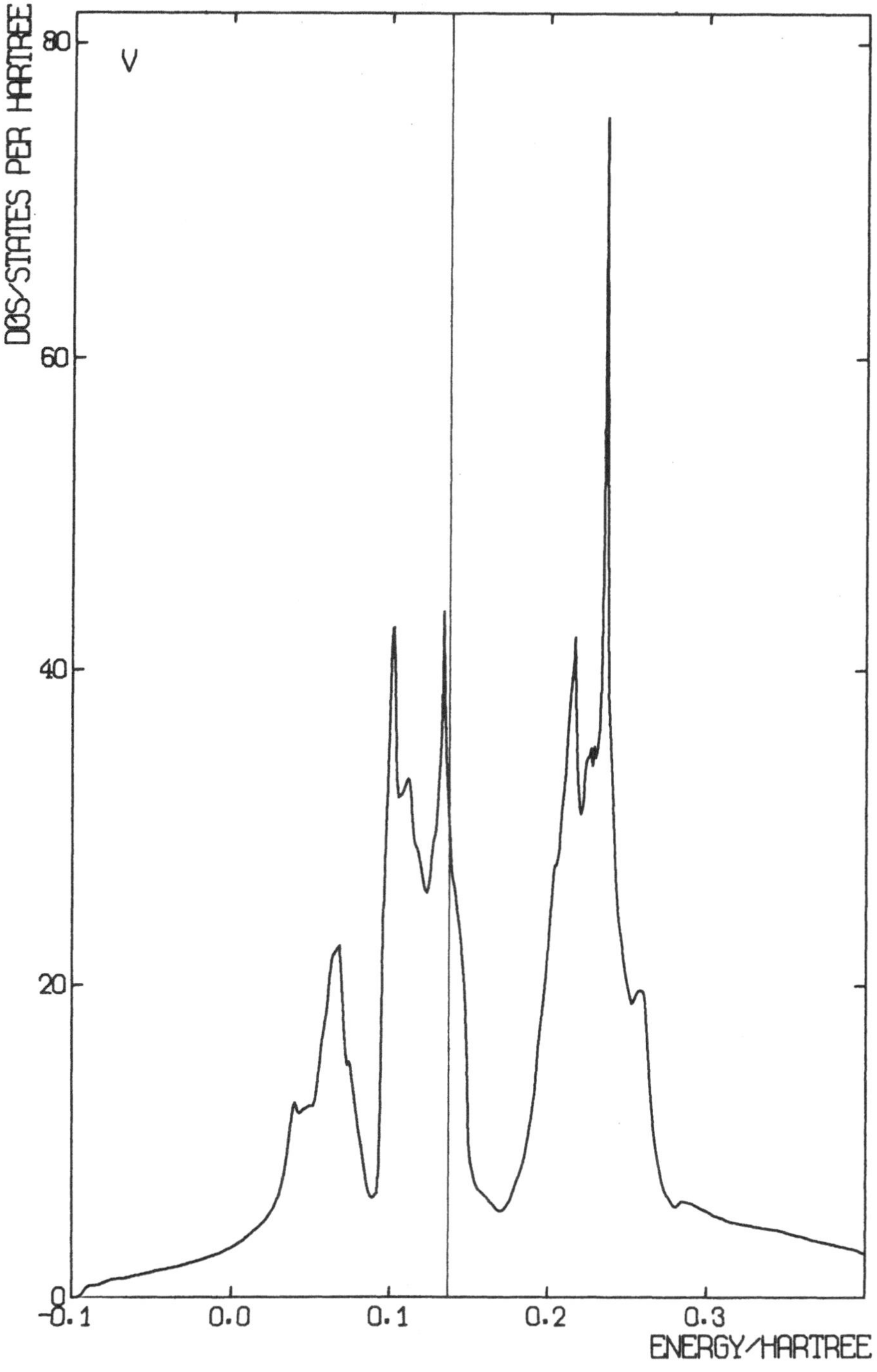

V
DOS/STATES PER HARTREE
0
20
40
60
80
-0.1
0.0
0.1
0.2
0.3
ENERGY/HARTREE

CHROMIUM, B.C.C. STRUCTURE, A = 5.44 A.U.

```
 EXPONENTS FOR THE POTENTIAL BASIS:
 24.0000      9.6981      3.9189      1.5836
 POTENTIAL EXPANSION COEFFICIENTS:
 -0.84702E+03 0.25837E+04-0.20222E+04 0.30957E+03
 -0.64359E+04 0.90560E+03 0.29038E+04-0.15409E+03
 -0.20471E+05 0.24788E+05-0.26115E+04 0.19705E+02

 EXPONENTS FOR THE ATOM-LIKE ORBITAL BASIS:
 24.0000      9.6981      3.9189      1.5836
 ORBITAL BASIS EXPANSION COEFFICIENTS:
   1S:
  0.26175E+03-0.42528E+02 0.10915E+02-0.57680E+00
  0.26954E+03 0.12115E+03-0.15878E+02 0.28391E+00
  0.28816E+04-0.35140E+03 0.91901E+01-0.35034E-01
   2S:
 -0.68927E+02 0.15648E+01-0.17277E+01 0.54290E+00
 -0.32074E+03 0.32230E+03 0.43580E+01-0.28057E+00
 -0.97755E+03 0.18080E+03-0.50759E+01 0.36254E-01
   2P:
 -0.15104E+03-0.25332E+03-0.13280E+02 0.32602E+00
 -0.98858E+03-0.10519E+03 0.18026E+02-0.14763E+00
 -0.55933E+04 0.17249E+03-0.83355E+01 0.16604E-01
   3S:
 -0.21276E+02-0.33824E+02 0.34969E+02-0.57425E+01
 -0.99865E+02 0.79618E+02-0.69137E+02 0.24911E+01
 -0.35015E+03-0.66224E+01 0.18567E+02-0.27799E+00
   3P:
 -0.48849E+01 0.18599E+03-0.26753E+02-0.22025E+01
 -0.88196E+02-0.10013E+03-0.30857E+01 0.82118E+00
 -0.79625E+03 0.64312E+03-0.79082E+01-0.77483E-01
   3D:
 -0.18360E+02 0.71749E+02 0.35231E+02 0.15302E+01
  0.25743E+03 0.58059E+01-0.12733E+02-0.22213E+00
 -0.31321E+04 0.14915E+03 0.56720E+01-0.30011E-02
   4S:
  0.10918E+03-0.41577E+03 0.39770E+03-0.10024E+03
  0.84974E+03-0.25331E+03-0.45905E+03 0.57871E+02
  0.29903E+04-0.35471E+04 0.56644E+03-0.58909E+01
   4P:
  0.51121E+01 0.20084E+02 0.64885E+02-0.30004E+02
  0.19296E+02-0.13998E+03-0.61220E+02 0.19861E+02
 -0.34524E+03-0.17402E+03 0.11746E+03-0.24370E+01

 EXPONENTS FOR THE DENSITY BASIS:
 48.0000     20.4570      8.7185      3.7157      1.5836
 SITE DENSITY EXPANSION COEFFICIENTS:
  0.22990E+06-0.14264E+06 0.36896E+05-0.58581E+03-0.22387E+02
  0.13814E+07 0.70384E+06-0.12029E+06 0.13072E+04 0.23016E+02
  0.26376E+08-0.32594E+07 0.11683E+06-0.47135E+03-0.34333E+01
 TOTAL ATOMIC SPHERE DENSITY EXPANSION COEFFICIENTS:
  0.22977E+06-0.14620E+06 0.44731E+05-0.64719E+04 0.16166E+04
  0.13834E+07 0.68268E+06-0.12385E+06 0.65043E+04-0.97201E+03
  0.26175E+08-0.33745E+07 0.16496E+06-0.69449E+04 0.16068E+03
```

SLATER-KOSTER INTEGRALS:

HAMILTONIAN INTEGRALS:

ONE-CENTRE:

(S)	(P0)	(P1-)	(P1+)	(D0)	(D1-)	(D1+)	(D2-)	(D2+)
0.1345	0.3140	0.3140	0.3140	0.1881	0.1829	0.1829	0.1829	0.1881

TWO-CENTRE:

(SSS)	(PSS)	(DSS)	(PPS)	(PPP)	(DPS)	(DPP)	(DDS)	(DDP)	(DDD)
-0.0176	0.0126	-0.0095	-0.0898	-0.0019	0.0005	-0.0120	-0.0172	0.0139	-0.0027
-0.0260	-0.0130	-0.0128	-0.0199	-0.0105	0.0109	-0.0088	-0.0107	0.0069	-0.0011
-0.0091	-0.0135	-0.0035	0.0209	-0.0040	0.0057	-0.0016	-0.0017	0.0007	-0.0001
-0.0027	-0.0039	-0.0009	0.0065	-0.0010	0.0014	-0.0004	-0.0003	0.0002	-0.0000
-0.0030	-0.0050	-0.0015	0.0090	-0.0002	0.0026	-0.0001	-0.0008	0.0000	0.0000

OVERLAP INTEGRALS:

ONE-CENTRE:

(S)	(P0)	(P1-)	(P1+)	(D0)	(D1-)	(D1+)	(D2-)	(D2+)
0.9502	0.9478	0.9478	0.9478	0.9992	0.9955	0.9955	0.9955	0.9992

TWO-CENTRE:

(SSS)	(PSS)	(DSS)	(PPS)	(PPP)	(DPS)	(DPP)	(DDS)	(DDP)	(DDD)
0.2148	0.2851	0.0866	-0.3866	0.1143	-0.1288	0.0654	0.0565	-0.0391	0.0065
0.1262	0.1817	0.0564	-0.2807	0.0507	-0.0939	0.0288	0.0340	-0.0161	0.0020
0.0119	0.0143	0.0042	-0.0248	0.0015	-0.0077	0.0010	0.0022	-0.0005	0.0001
0.0023	0.0007	0.0004	0.0004	-0.0003	-0.0002	0.0000	0.0001	-0.0001	0.0000
0.0023	0.0016	0.0009	-0.0014	-0.0007	-0.0013	-0.0002	0.0006	0.0001	-0.0000

FERMI ENERGY = 0.1987 HARTREE

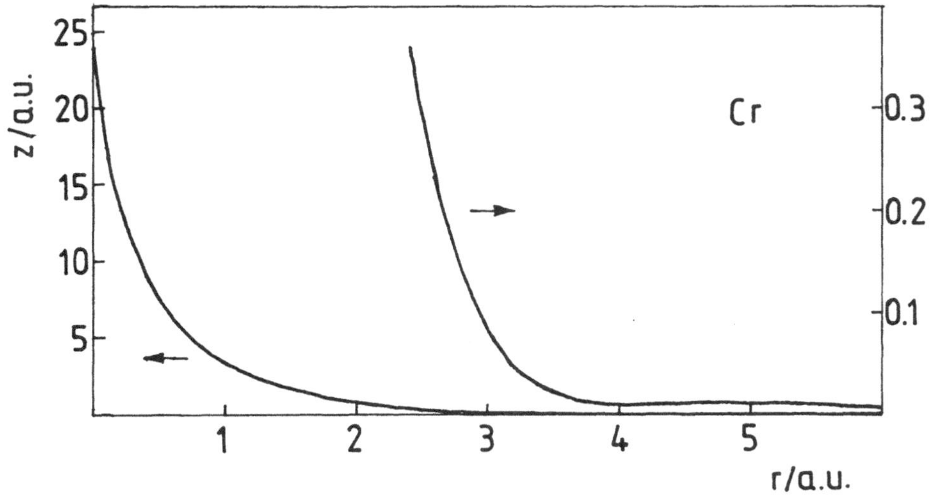

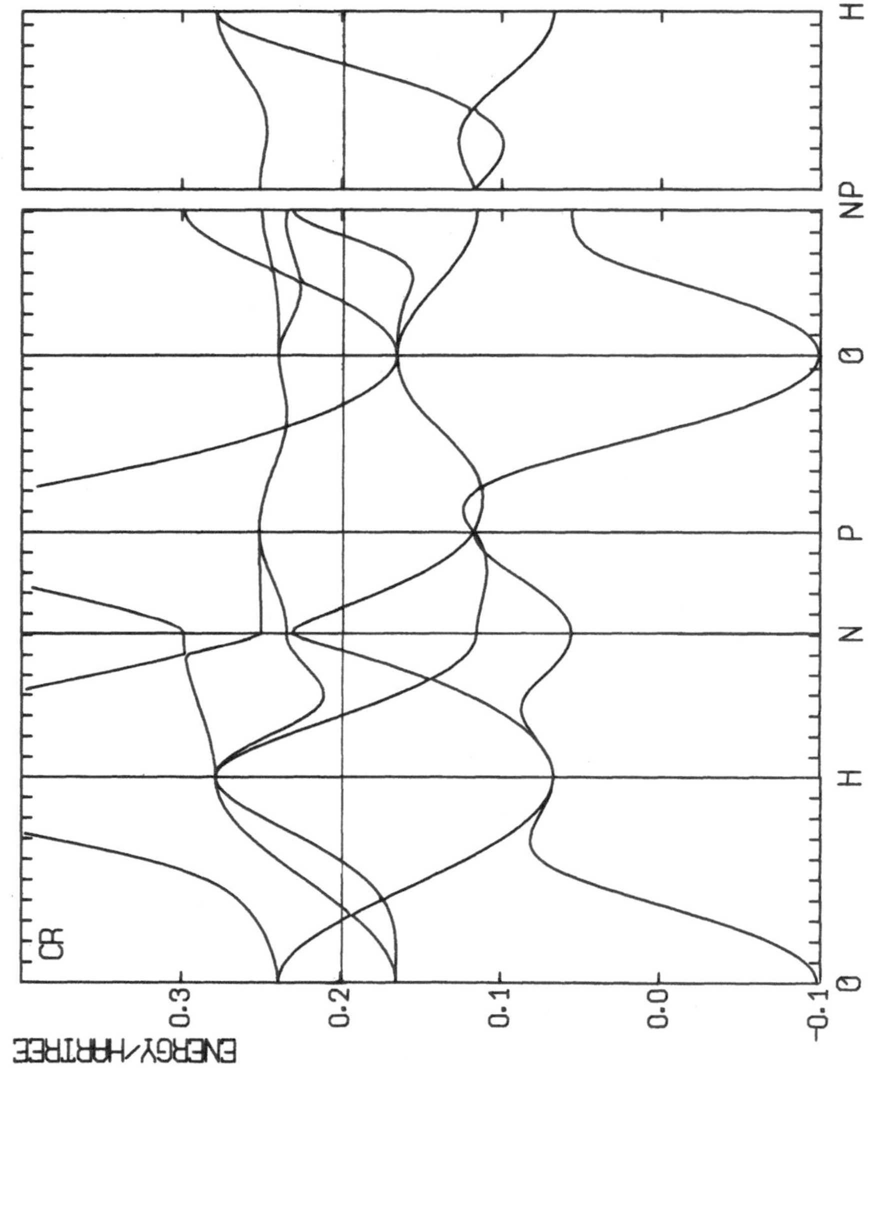
ENERGY/HARTREE
0.3
0.2
0.1
0.0
-0.1
CR
Ø
H
N
P
Ø
NP
H

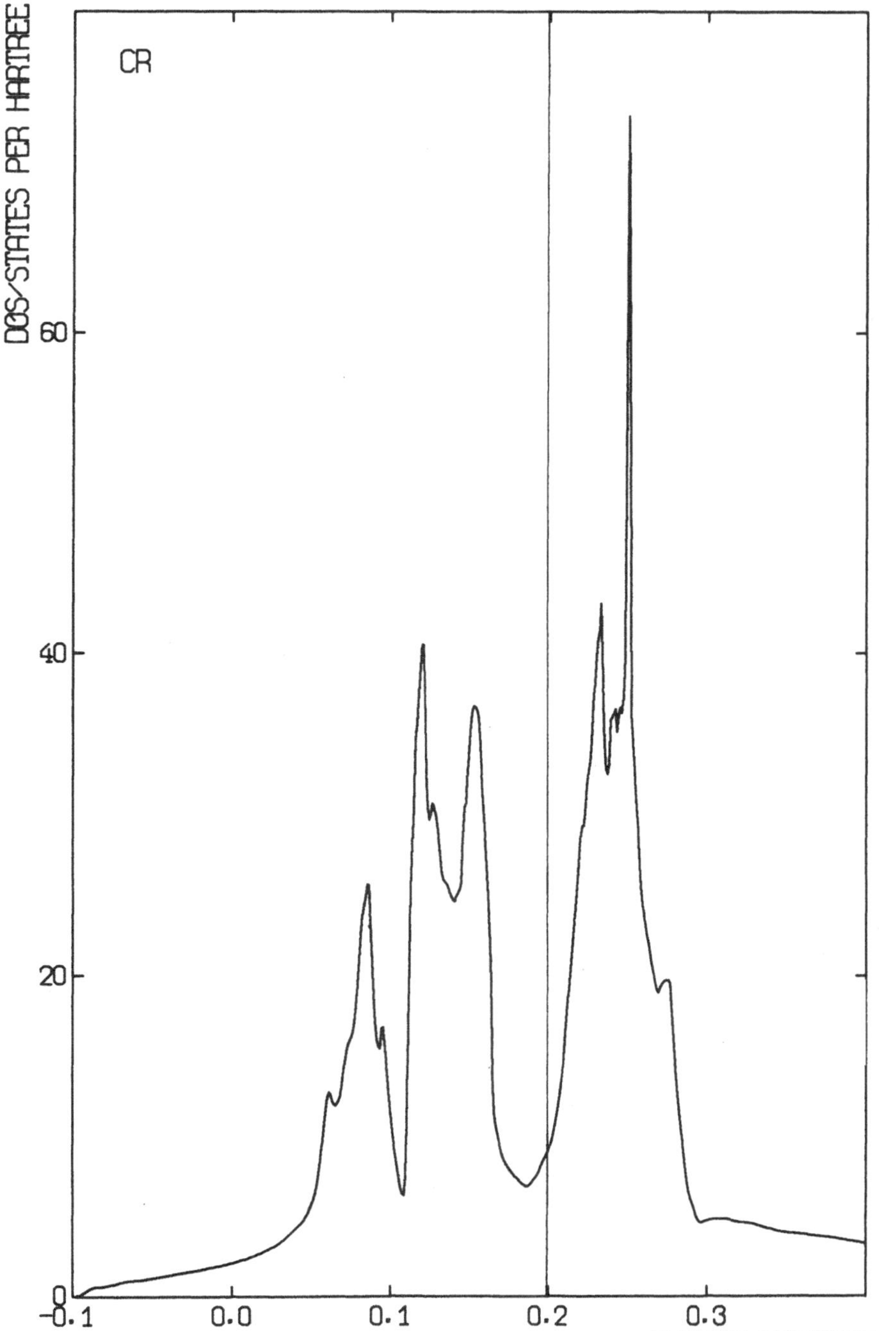
CR
DOS/STATES PER HARTREE
ENERGY/HARTREE
0
20
40
60
-0.1
0.0
0.1
0.2
0.3

```
MANGANESE, F.C.C. STRUCTURE, A = 6.95 A.U.

 EXPONENTS FOR THE POTENTIAL BASIS:
 25.0000       9.9212       3.9372       1.5625
 POTENTIAL EXPANSION COEFFICIENTS:
  0.32576E+03-0.94596E+03 0.78349E+03-0.13829E+03
  0.26248E+04-0.21221E+03-0.10018E+04 0.95650E+02
  0.94414E+04-0.10023E+05 0.10410E+04-0.15642E+02

 EXPONENTS FOR THE ATOM-LIKE ORBITAL BASIS:
 25.0000       9.9212       3.9372       1.5625
 ORBITAL BASIS EXPANSION COEFFICIENTS:
   1S:
  0.25403E+03-0.16187E+02 0.46655E+01-0.26146E+00
  0.15719E+03 0.45915E+02-0.70029E+01 0.13055E+00
  0.16224E+04-0.15248E+03 0.42239E+01-0.16309E-01
   2S:
 -0.63715E+02-0.19368E+02 0.10712E+02-0.47870E+00
 -0.30388E+03 0.38768E+03-0.15113E+02 0.23828E+00
 -0.92349E+03-0.14224E+03 0.83659E+01-0.29697E-01
   2P:
  0.71471E+02 0.40026E+03-0.45072E+01 0.13487E+00
  0.31092E+03-0.35382E+03 0.67642E+01-0.61741E-01
  0.84404E+02 0.70171E+03-0.34084E+01 0.70035E-02
   3S:
 -0.29332E+02-0.13160E+01 0.51484E+01-0.21899E+01
 -0.16734E+03 0.15824E+03-0.32089E+02 0.85712E+00
 -0.51781E+03 0.33276E+03-0.10165E+02-0.85195E-01
   3P:
 -0.33650E+02-0.17037E+03 0.28884E+02 0.13091E+01
 -0.19336E+03-0.55189E+02 0.31998E+01-0.47522E+00
 -0.75828E+03-0.29923E+03 0.10144E+02 0.43194E-01
   3D:
 -0.78009E+02 0.17270E+03 0.24755E+02 0.15140E+01
 -0.47223E+03-0.29873E+03 0.39519E+01-0.30572E+00
 -0.56849E+04 0.10574E+04-0.66319E+00 0.10900E-01
   4S:
 -0.90952E+02 0.34247E+03-0.32459E+03 0.82341E+02
 -0.75516E+03 0.17243E+03 0.38879E+03-0.48622E+02
 -0.29142E+04 0.31626E+04-0.48521E+03 0.49259E+01
   4P:
  0.21066E+02 0.70796E+01 0.60696E+02-0.25787E+02
  0.14405E+03-0.72718E+02-0.66933E+02 0.17383E+02
  0.30778E+03-0.38810E+03 0.11375E+03-0.21343E+01

 EXPONENTS FOR THE DENSITY BASIS:
 50.0000      21.0223       8.8388       3.7162       1.5625
 SITE DENSITY EXPANSION COEFFICIENTS:
  0.11529E+06 0.69110E+04 0.48416E+04 0.80523E+03-0.29716E+02
 -0.27062E+05-0.20958E+06-0.30320E+05-0.31126E+03 0.24323E+02
 -0.37071E+07 0.17894E+07 0.22079E+05 0.65715E+02-0.34144E+01
 TOTAL ATOMIC SPHERE DENSITY EXPANSION COEFFICIENTS:
  0.11565E+06 0.39206E+04 0.10294E+05-0.32471E+04 0.11447E+04
 -0.18983E+05-0.22102E+06-0.33343E+05 0.32105E+04-0.69984E+03
 -0.37653E+07 0.16821E+07 0.57001E+05-0.45504E+04 0.11817E+03
```

SLATER-KOSTER INTEGRALS:

HAMILTONIAN INTEGRALS:

ONE-CENTRE:

(S)	(P0)	(P1-)	(P1+)	(D0)	(D1-)	(D1+)	(D2-)	(D2+)
0.1334	0.3171	0.3171	0.3171	0.1840	0.1812	0.1812	0.1812	0.1840

TWO-CENTRE:

(SSS)	(PSS)	(DSS)	(PPS)	(PPP)	(DPS)	(DPP)	(DDS)	(DDP)	(DDD)
-0.0285	0.0023	0.0144	-0.0566	-0.0086	0.0101	-0.0125	-0.0147	0.0106	-0.0018
-0.0156	0.0216	0.0062	0.0282	-0.0062	0.0096	-0.0024	-0.0031	0.0010	-0.0001
-0.0046	0.0070	0.0014	0.0119	-0.0018	0.0023	-0.0006	-0.0005	0.0002	-0.0000
-0.0018	0.0029	0.0008	0.0054	-0.0001	0.0013	-0.0000	-0.0003	0.0000	0.0000

OVERLAP INTEGRALS:

ONE-CENTRE:

(S)	(P0)	(P1-)	(P1+)	(D0)	(D1-)	(D1+)	(D2-)	(D2+)
0.9687	0.9648	0.9648	0.9648	0.9991	0.9982	0.9982	0.9982	0.9991

TWO-CENTRE:

(SSS)	(PSS)	(DSS)	(PPS)	(PPP)	(DPS)	(DPP)	(DDS)	(DDP)	(DDD)
0.1899	-0.2608	-0.0697	-0.3692	0.0968	-0.1086	0.0478	0.0399	-0.0242	0.0037
0.0301	-0.0440	-0.0112	-0.0789	0.0061	-0.0211	0.0031	0.0054	-0.0013	0.0001
0.0045	-0.0036	-0.0011	-0.0051	-0.0002	-0.0016	0.0001	0.0004	-0.0001	0.0000
0.0011	-0.0001	-0.0004	0.0009	-0.0006	-0.0003	-0.0001	0.0002	0.0000	-0.0000

FERMI ENERGY = 0.2000 HARTREE

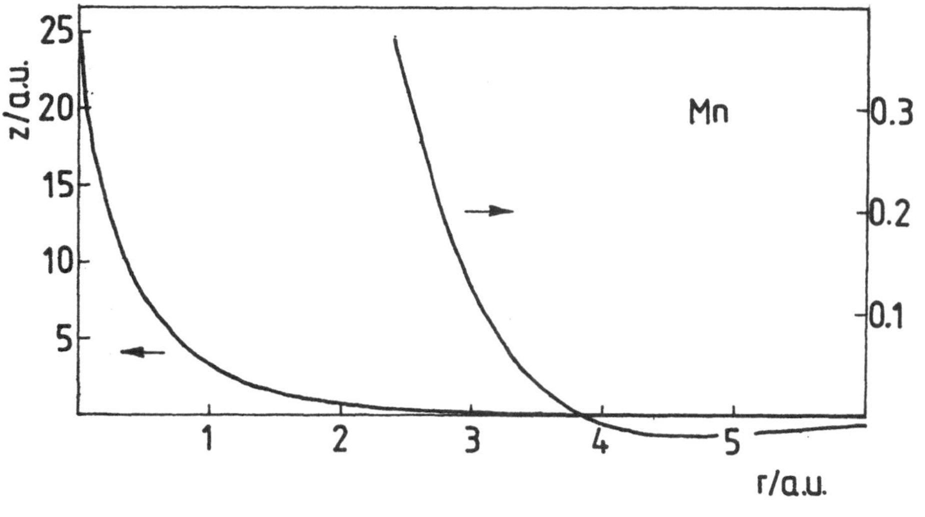

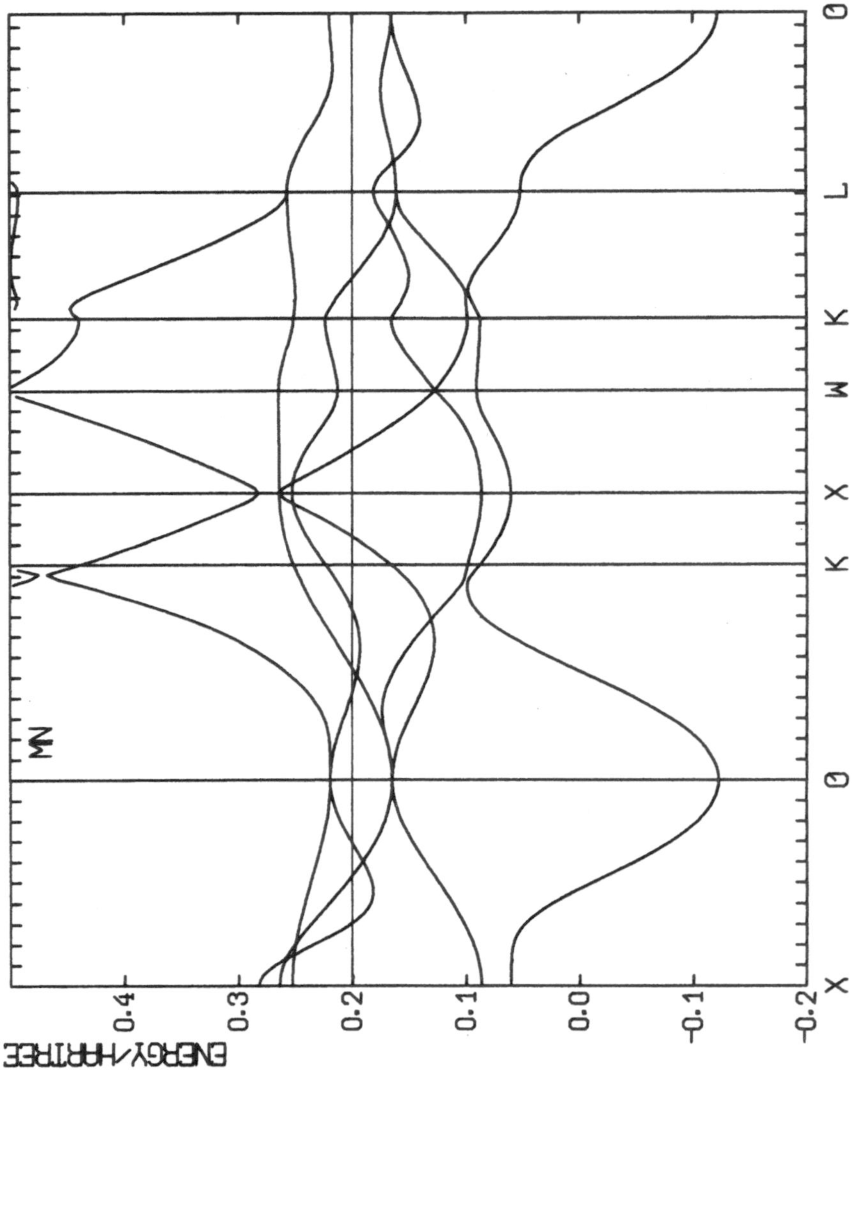
MN
ENERGY/HARTREE
-0.2
-0.1
0.0
0.1
0.2
0.3
0.4
X
0
K
X
W
K
L
0

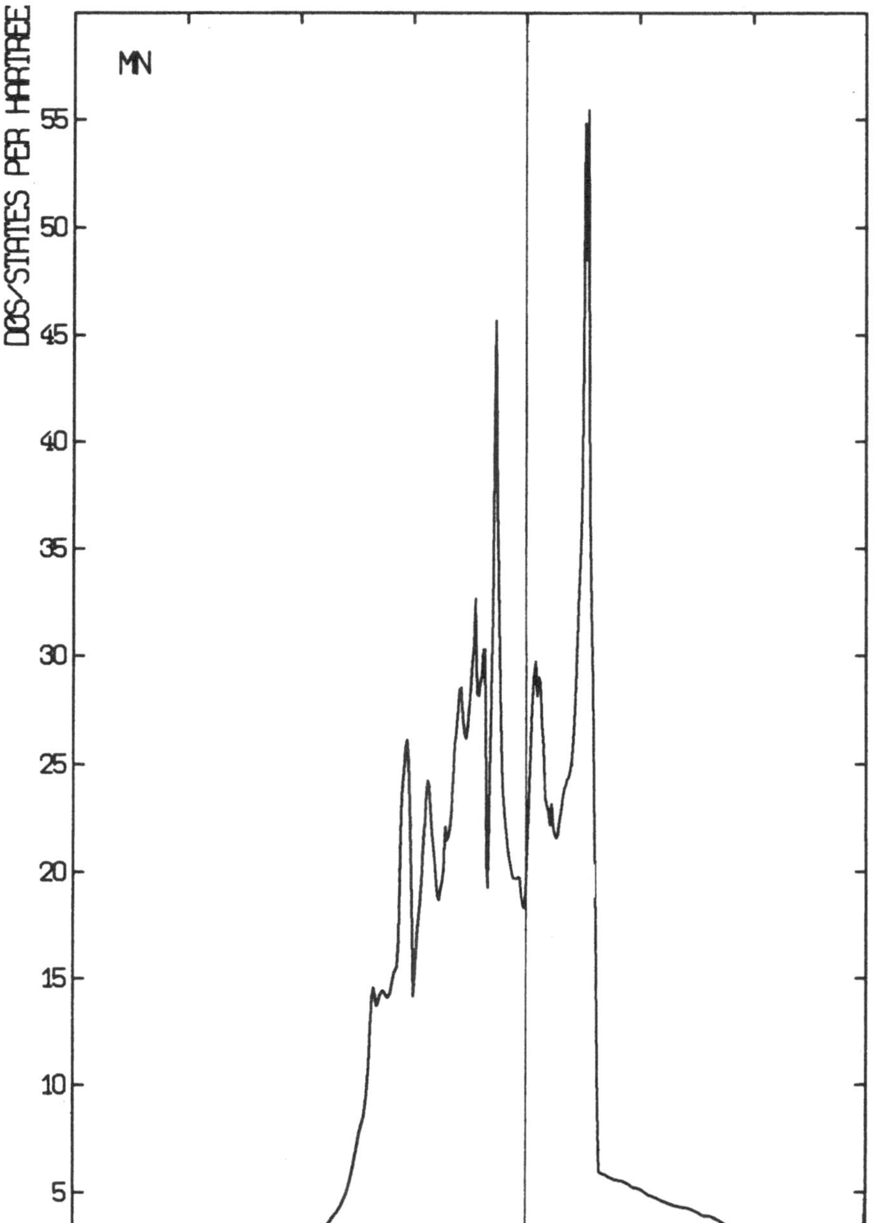
MN
DOS/STATES PER HARTREE
55
50
45
40
35
30
25
20
15
10
5
0
-0.2
-0.1
0.0
0.1
0.2
0.3
0.4
ENERGY/HARTREE

```
IRON, B.C.C. STRUCTURE, A = 5.41 A.U.

 EXPONENTS FOR THE POTENTIAL BASIS:
 26.0000     10.2479     4.0392      1.5920
 POTENTIAL EXPANSION COEFFICIENTS:
  0.64602E+03-0.18613E+04 0.14851E+04-0.24385E+03
  0.54674E+04-0.35595E+03-0.20807E+04 0.15839E+03
  0.20447E+05-0.21329E+05 0.21018E+04-0.25150E+02

 EXPONENTS FOR THE ATOM-LIKE ORBITAL BASIS:
 26.0000     10.2479     4.0392      1.5920
 ORBITAL BASIS EXPANSION COEFFICIENTS:
   1S:
  0.26258E+03-0.90406E+01 0.30870E+01-0.18345E+00
  0.13412E+03 0.24361E+02-0.48204E+01 0.94512E-01
  0.12563E+04-0.99437E+02 0.30653E+01-0.12173E-01
   2S:
  0.64680E+02 0.26589E+02-0.14457E+02 0.76754E+00
  0.32051E+03-0.44699E+03 0.21704E+02-0.39590E+00
  0.98252E+03 0.29639E+03-0.13195E+02 0.51070E-01
   2P:
 -0.52593E+02-0.47829E+03 0.98787E+01-0.26520E+00
 -0.12556E+03 0.56371E+03-0.14806E+02 0.12444E+00
  0.21994E+04-0.10863E+04 0.74418E+01-0.14463E-01
   3S:
  0.32553E+02-0.85987E+01 0.44000E+01 0.12372E+01
  0.19923E+03-0.20675E+03 0.22222E+02-0.42215E+00
  0.61824E+03-0.47961E+03 0.19281E+02 0.33630E-01
   3P:
 -0.49170E+02-0.18068E+03 0.33139E+02 0.10691E+01
 -0.31857E+03-0.11465E+03 0.12104E+01-0.38839E+00
 -0.15027E+04-0.19441E+03 0.12427E+02 0.35022E-01
   3D:
 -0.10782E+03 0.23163E+03 0.24568E+02 0.15970E+01
 -0.84216E+03-0.45732E+03 0.87057E+01-0.35346E+00
 -0.78493E+04 0.16111E+04-0.25938E+01 0.15890E-01
   4S:
  0.88907E+02-0.33216E+03 0.31322E+03-0.79594E+02
  0.77399E+03-0.15237E+03-0.39026E+03 0.48517E+02
  0.31674E+04-0.33178E+04 0.50109E+03-0.50246E+01
   4P:
  0.27036E+02 0.49100E+01 0.62009E+02-0.25867E+02
  0.19811E+03-0.57104E+02-0.73118E+02 0.17981E+02
  0.58447E+03-0.50942E+03 0.12475E+03-0.22620E+01

 EXPONENTS FOR THE DENSITY BASIS:
 52.0000     21.7516     9.0987      3.8060      1.5920
 SITE DENSITY EXPANSION COEFFICIENTS:
  0.13705E+06 0.30079E+04 0.37452E+04 0.88502E+03-0.29063E+02
  0.48662E+05-0.17140E+06-0.30951E+05-0.39529E+03 0.24459E+02
 -0.18191E+07 0.21742E+07 0.27134E+05 0.10305E+03-0.34992E+01
 TOTAL ATOMIC SPHERE DENSITY EXPANSION COEFFICIENTS:
  0.14131E+06-0.65452E+04 0.13195E+05-0.45852E+04 0.13868E+04
  0.11338E+06-0.16293E+06-0.42105E+05 0.48782E+04-0.85949E+03
 -0.10093E+07 0.17945E+07 0.89050E+05-0.62368E+04 0.14618E+03
```

SLATER-KOSTER INTEGRALS:

HAMILTONIAN INTEGRALS:

ONE-CENTRE:

(S)	(P0)	(P1-)	(P1+)	(D0)	(D1-)	(D1+)	(D2-)	(D2+)
0.1467	0.3405	0.3405	0.3405	0.1940	0.1907	0.1907	0.1907	0.1940

TWO-CENTRE:

(SSS)	(PSS)	(DSS)	(PPS)	(PPP)	(DPS)	(DPP)	(DDS)	(DDP)	(DDD)
-0.0309	-0.0006	-0.0149	-0.0684	-0.0087	0.0102	-0.0141	-0.0159	0.0117	-0.0021
-0.0310	-0.0273	-0.0138	-0.0033	-0.0130	0.0149	-0.0088	-0.0091	0.0054	-0.0008
-0.0087	-0.0132	-0.0028	0.0211	-0.0035	0.0047	-0.0011	-0.0011	0.0004	-0.0001
-0.0023	-0.0034	-0.0006	0.0060	-0.0007	0.0010	-0.0002	-0.0002	0.0001	-0.0000
-0.0024	-0.0040	-0.0010	0.0075	-0.0001	0.0016	-0.0000	-0.0004	0.0000	0.0000

OVERLAP INTEGRALS:

ONE-CENTRE:

(S)	(P0)	(P1-)	(P1+)	(D0)	(D1-)	(D1+)	(D2-)	(D2+)
0.9732	0.9695	0.9695	0.9695	0.9998	0.9983	0.9983	0.9983	0.9998

TWO-CENTRE:

(SSS)	(PSS)	(DSS)	(PPS)	(PPP)	(DPS)	(DPP)	(DDS)	(DDP)	(DDD)
0.2060	0.2796	0.0702	-0.3871	0.1105	-0.1075	0.0514	0.0391	-0.0250	0.0040
0.1184	0.1736	0.0435	-0.2731	0.0487	-0.0742	0.0224	0.0218	-0.0099	0.0012
0.0106	0.0128	0.0032	-0.0229	0.0008	-0.0057	0.0006	0.0012	-0.0003	0.0000
0.0018	0.0002	0.0003	0.0011	-0.0005	-0.0001	-0.0001	0.0001	-0.0000	0.0000
0.0014	0.0003	0.0005	0.0007	-0.0007	-0.0004	-0.0001	0.0002	0.0000	-0.0000

FERMI ENERGY = 0.2319 HARTREE

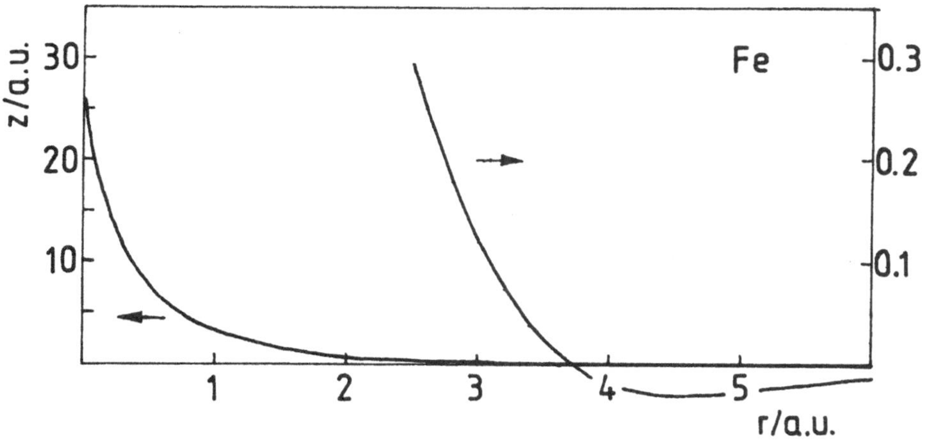

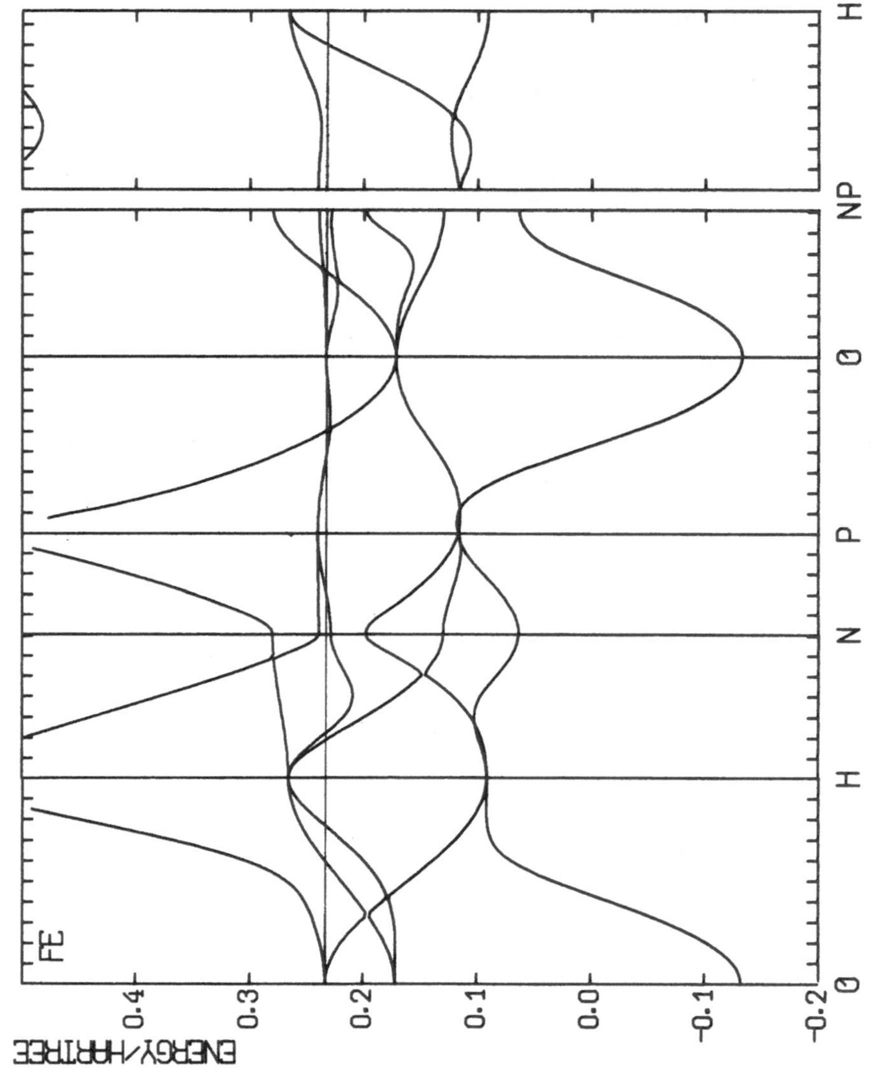
FE
ENERGY/HARTREE
0.4
0.3
0.2
0.1
0.0
-0.1
-0.2
H
N
P
NP
H

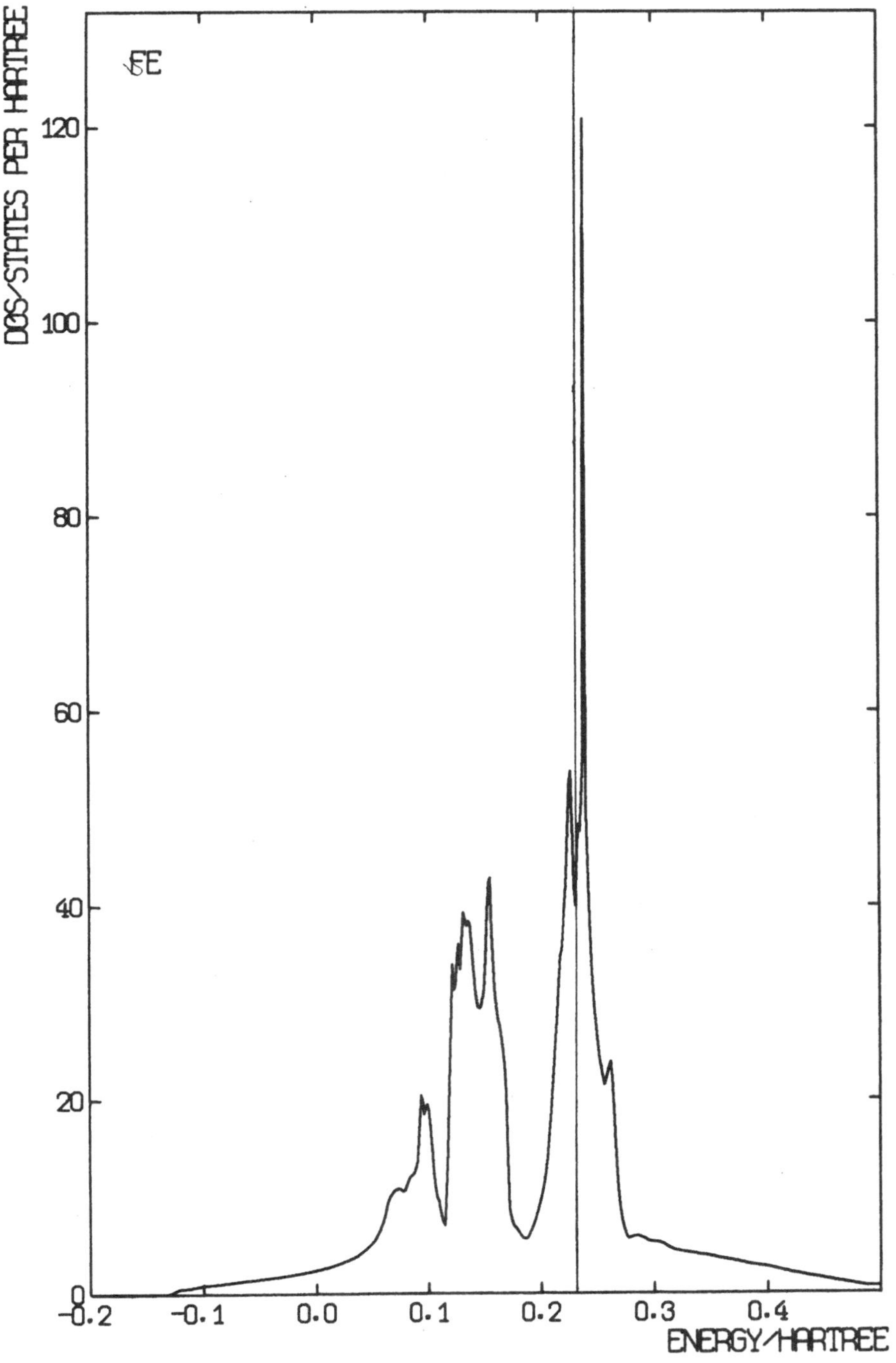
FE
DOS/STATES PER HARTREE
0
20
40
60
80
100
120
-0.2
-0.1
0.0
0.1
0.2
0.3
0.4
ENERGY/HARTREE

```
COBALT, H.C.P. STRUCTURE, A = 4.73 A.U., C = 7.68 A.U.

 EXPONENTS FOR THE POTENTIAL BASIS:
 27.0000     10.5791      4.1451      1.6241
 POTENTIAL EXPANSION COEFFICIENTS:
  0.60221E+02-0.80462E+01-0.12396E+03 0.98786E+02
  0.55537E+03 0.52568E+03 0.39763E+02-0.60020E+02
  0.27738E+04-0.94353E+03-0.35211E+03 0.11985E+02

 EXPONENTS FOR THE ATOM-LIKE ORBITAL BASIS:
 27.0000     10.5791      4.1451      1.6241
 ORBITAL BASIS EXPANSION COEFFICIENTS:
   1S:
 -0.28831E+03 0.20772E+02-0.54889E+01 0.29721E+00
 -0.20053E+03-0.66712E+02 0.87747E+01-0.15641E+00
 -0.23742E+04 0.22111E+03-0.55643E+01 0.20566E-01
   2S:
  0.73887E+02 0.17131E+02-0.84044E+01 0.32402E+00
  0.38861E+03-0.49140E+03 0.12559E+02-0.16831E+00
  0.12689E+04 0.23283E+03-0.70542E+01 0.21859E-01
   2P:
  0.11483E+03 0.46695E+03-0.24642E+01 0.66152E-01
  0.68578E+03-0.42525E+03 0.38834E+01-0.31776E-01
  0.22755E+04 0.69476E+03-0.20179E+01 0.37781E-02
   3S:
 -0.32027E+02-0.73420E+00 0.32954E+01-0.21920E+01
 -0.19778E+03 0.22710E+03-0.39953E+02 0.97758E+00
 -0.68469E+03 0.39147E+03-0.55288E+01-0.11133E+00
   3P:
 -0.35738E+02-0.22416E+03 0.41918E+02 0.10594E+01
 -0.21893E+03-0.25048E+02-0.72884E+01-0.40860E+00
 -0.79352E+03-0.46662E+03 0.15928E+02 0.39460E-01
   3D:
 -0.80509E+02 0.21046E+03 0.40352E+02 0.15166E+01
 -0.35611E+03-0.29197E+03-0.10677E+02-0.30999E+00
 -0.80845E+04 0.12842E+04 0.60049E+01 0.10156E-01
   4S:
 -0.82634E+02 0.31055E+03-0.29441E+03 0.76555E+02
 -0.75500E+03 0.13020E+03 0.37531E+03-0.48400E+02
 -0.32430E+04 0.33414E+04-0.50370E+03 0.51474E+01
   4P:
  0.30100E+02 0.95062E+01 0.59432E+02-0.25457E+02
  0.23386E+03-0.48879E+02-0.72801E+02 0.18401E+02
  0.90888E+03-0.56955E+03 0.12928E+03-0.23793E+01

 EXPONENTS FOR THE DENSITY BASIS:
 54.0000     22.4880      9.3650      3.9000      1.6241
 SITE DENSITY EXPANSION COEFFICIENTS:
  0.26498E+06-0.12713E+06 0.35383E+05-0.17488E+04 0.10222E+03
  0.14539E+07 0.68093E+06-0.13517E+06 0.29529E+04-0.44117E+02
  0.30728E+08-0.26211E+07 0.18423E+06-0.15987E+04 0.53891E+01
 TOTAL ATOMIC SPHERE DENSITY EXPANSION COEFFICIENTS:
  0.34623E+06-0.25462E+06 0.10294E+06-0.22444E+05 0.33352E+04
  0.26441E+07 0.11708E+07-0.28941E+06 0.27979E+05-0.19787E+04
  0.51425E+08-0.81166E+07 0.61239E+06-0.23017E+05 0.31409E+03
```

SLATER-KOSTER INTEGRALS:

HAMILTONIAN INTEGRALS:

ONE-CENTRE:

(S)	(P0)	(P1-)	(P1+)	(D0)	(D1-)	(D1+)	(D2-)	(D2+)
-0.0374	0.1682	0.1691	0.1691	-0.0019	-0.0006	-0.0006	-0.0011	-0.0011

TWO-CENTRE:

(SSS)	(PSS)	(DSS)	(PPS)	(PPP)	(DPS)	(DPP)	(DDS)	(DDP)	(DDD)
-0.0548	0.0427	0.0238	-0.0069	-0.0196	0.0249	-0.0169	-0.0183	0.0117	-0.0019
-0.0549	0.0420	0.0239	-0.0082	-0.0196	0.0249	-0.0171	-0.0186	0.0120	-0.0020
-0.0207	0.0298	0.0077	0.0424	-0.0071	0.0124	-0.0027	-0.0035	0.0011	-0.0001
-0.0085	0.0128	0.0027	0.0211	-0.0030	0.0045	-0.0009	-0.0010	0.0003	-0.0000
-0.0049	0.0074	0.0015	0.0127	-0.0017	0.0024	-0.0005	-0.0005	0.0002	-0.0000
-0.0049	0.0075	0.0015	0.0129	-0.0017	0.0024	-0.0005	-0.0005	0.0002	-0.0000
-0.0021	0.0032	0.0007	0.0059	-0.0004	0.0011	-0.0001	-0.0002	0.0001	-0.0000
-0.0017	0.0026	0.0006	0.0048	-0.0000	0.0011	-0.0000	-0.0003	0.0000	0.0000

OVERLAP INTEGRALS:

ONE-CENTRE:

(S)	(P0)	(P1-)	(P1+)	(D0)	(D1-)	(D1+)	(D2-)	(D2+)
0.9774	0.9734	0.9741	0.9741	0.9989	0.9992	0.9992	0.9991	0.9991

TWO-CENTRE:

(SSS)	(PSS)	(DSS)	(PPS)	(PPP)	(DPS)	(DPP)	(DDS)	(DDP)	(DDD)
0.1831	-0.2546	-0.0634	-0.3646	0.0934	-0.0998	0.0429	0.0338	-0.0203	0.0031
0.1855	-0.2572	-0.0641	-0.3671	0.0951	-0.1005	0.0437	0.0342	-0.0207	0.0031
0.0284	-0.0417	-0.0099	-0.0755	0.0057	-0.0189	0.0027	0.0045	-0.0011	0.0001
0.0078	-0.0086	-0.0022	-0.0152	0.0002	-0.0039	0.0003	0.0009	-0.0002	0.0000
0.0040	-0.0030	-0.0009	-0.0043	-0.0004	-0.0013	0.0000	0.0003	-0.0001	0.0000
0.0040	-0.0031	-0.0009	-0.0044	-0.0004	-0.0014	0.0000	0.0003	-0.0001	0.0000
0.0013	0.0001	-0.0002	0.0016	-0.0006	-0.0000	-0.0001	0.0001	0.0000	-0.0000
0.0008	0.0003	-0.0002	0.0015	-0.0006	-0.0001	-0.0001	0.0001	0.0000	-0.0000

FERMI ENERGY = 0.0485 HARTREE

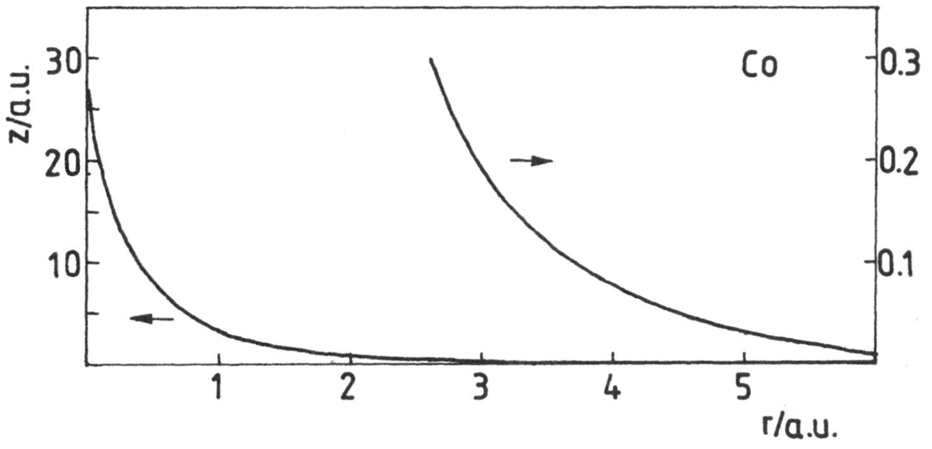

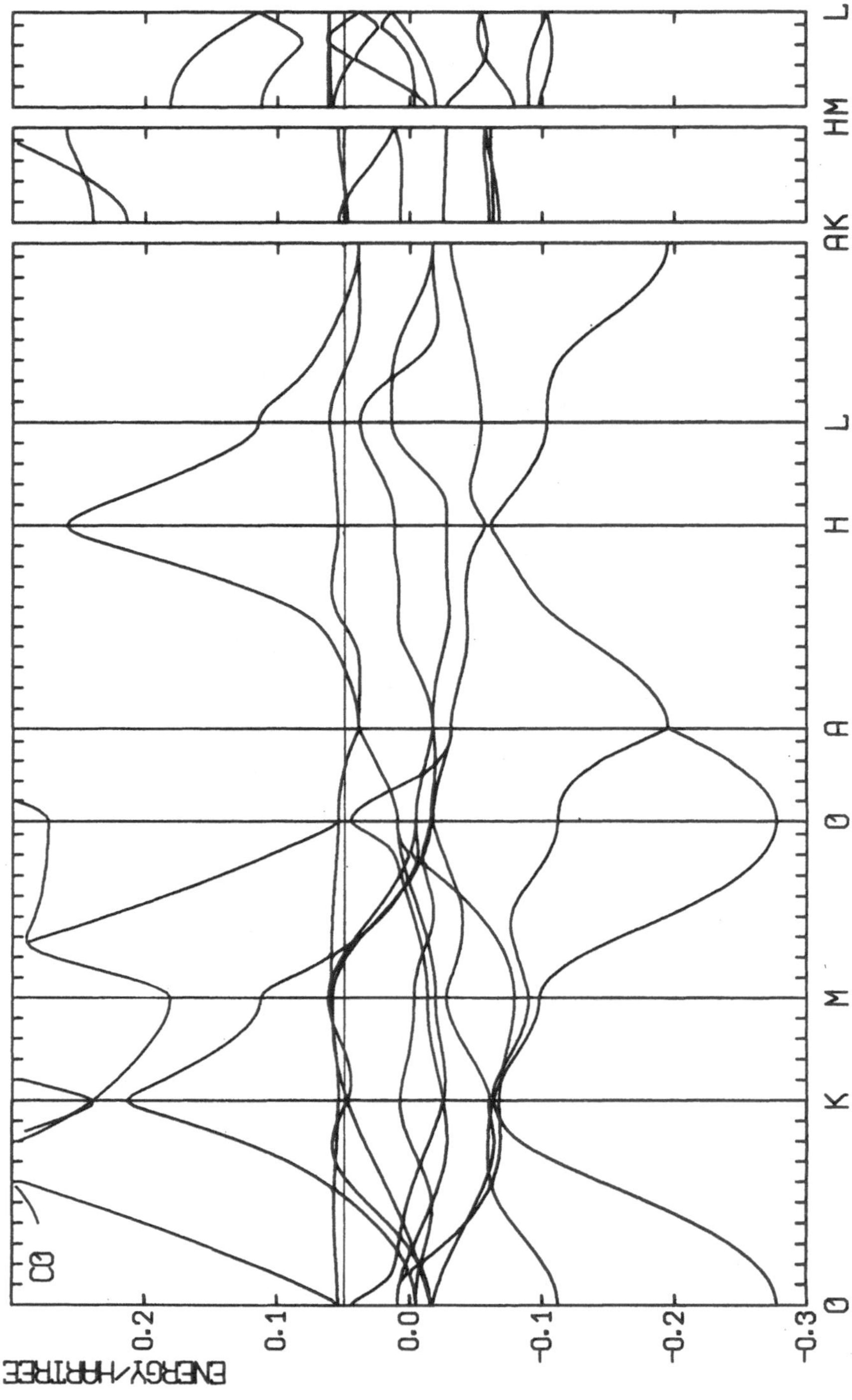
CO
ENERGY/HARTREE
0.2
0.1
0.0
-0.1
-0.2
-0.3
0
K
M
0
A
H
L
AK
HM
L

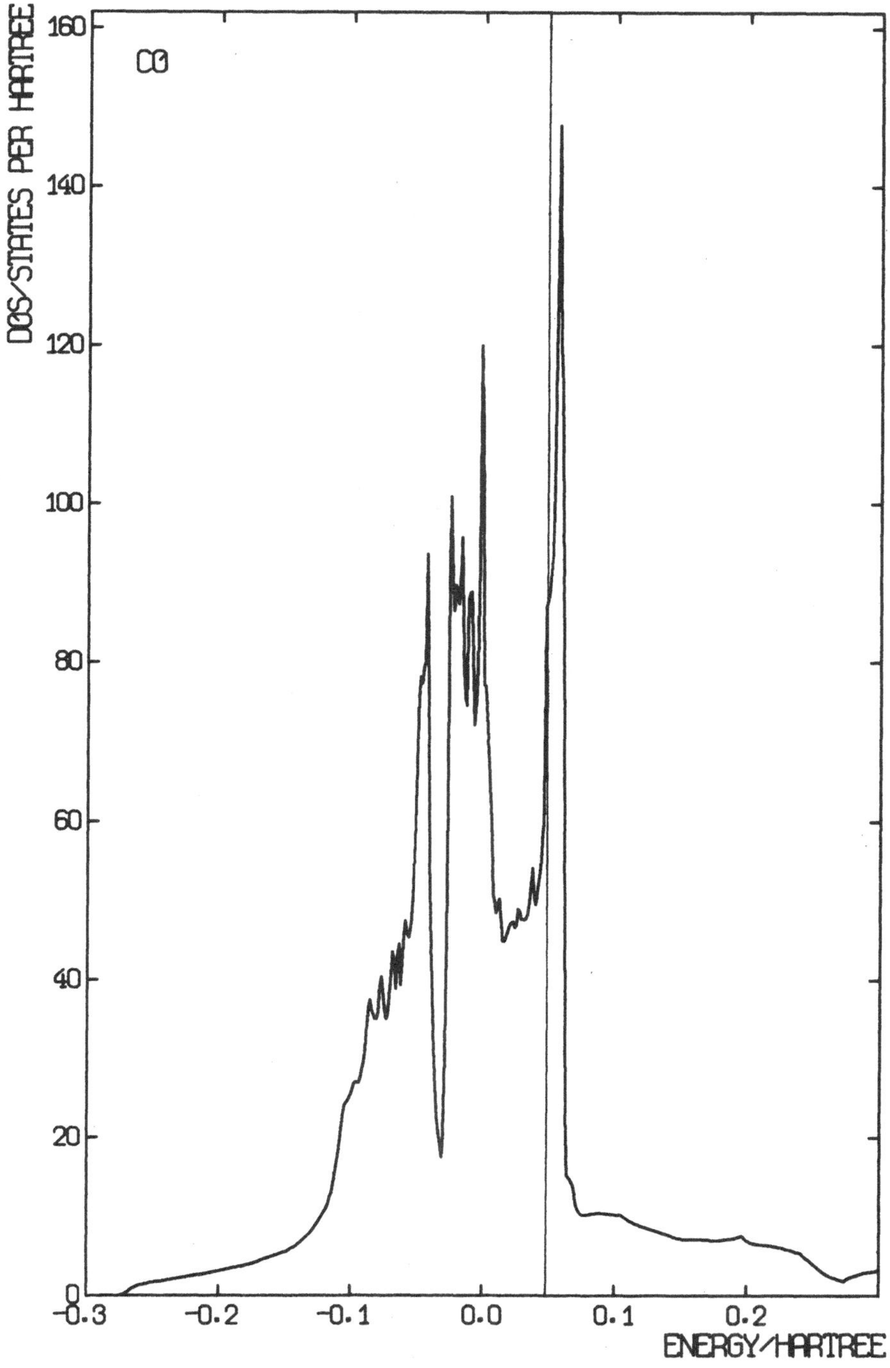
DOS/STATES PER HARTREE
160
140
120
100
80
60
40
20
0
CO
-0.3
-0.2
-0.1
0.0
0.1
0.2
ENERGY/HARTREE

```
NICKEL, F.C.C. STRUCTURE, A = 6.65 A.U.

 EXPONENTS FOR THE POTENTIAL BASIS:
 28.0000      10.8526      4.2064      1.6304
 POTENTIAL EXPANSION COEFFICIENTS:
 -0.55404E+03 0.16349E+04-0.12836E+04 0.23078E+03
 -0.51710E+04 0.38710E+03 0.20207E+04-0.12836E+03
 -0.20870E+05 0.21138E+05-0.21631E+04 0.18569E+02

 EXPONENTS FOR THE ATOM-LIKE ORBITAL BASIS:
 28.0000      10.8526      4.2064      1.6304
 ORBITAL BASIS EXPANSION COEFFICIENTS:
   1S:
  0.31587E+03-0.33306E+02 0.77116E+01-0.39509E+00
  0.26457E+03 0.11716E+03-0.12598E+02 0.21077E+00
  0.39515E+04-0.36248E+03 0.80242E+01-0.28062E-01
   2S:
 -0.87965E+02-0.10579E+01 0.17384E+00 0.23957E+00
 -0.50230E+03 0.51640E+03 0.81387E+00-0.13596E+00
 -0.18170E+04-0.95794E+02-0.22885E+01 0.19170E-01
   2P:
  0.19847E+03 0.43729E+03 0.62779E+01-0.15836E+00
  0.14629E+04-0.21354E+03-0.97513E+01 0.77635E-01
  0.10238E+05 0.85319E+02 0.50790E+01-0.94109E-02
   3S:
  0.29046E+02 0.14949E+02-0.13146E+02 0.30125E+01
  0.17099E+03-0.25917E+03 0.63163E+02-0.14884E+01
  0.66695E+03-0.18002E+03-0.13649E+02 0.18666E+00
   3P:
  0.24719E+02 0.26300E+03-0.47023E+02-0.10679E+01
  0.13173E+03-0.88520E+02 0.10499E+02 0.44521E+00
  0.15207E+03 0.67868E+03-0.14992E+02-0.46780E-01
   3D:
 -0.88731E+01 0.14679E+03 0.60218E+02 0.12132E+01
  0.55325E+03 0.10091E+03-0.38235E+02-0.18395E+00
 -0.40035E+04 0.45229E+03 0.18834E+02-0.26133E-02
   4S:
  0.73023E+02-0.27361E+03 0.25731E+03-0.66967E+02
  0.70462E+03-0.86340E+02-0.34007E+03 0.43175E+02
  0.32051E+04-0.31793E+04 0.46423E+03-0.44409E+01
   4P:
  0.58167E+02-0.27754E+02 0.71775E+02-0.24922E+02
  0.49207E+03 0.48312E+02-0.99602E+02 0.18163E+02
  0.31580E+04-0.11400E+04 0.14929E+03-0.23517E+01

 EXPONENTS FOR THE DENSITY BASIS:
 56.0000      23.1319      9.5551      3.9469      1.6304
 SITE DENSITY EXPANSION COEFFICIENTS:
  0.34093E+06-0.18654E+06 0.45368E+05-0.19008E+04 0.39581E+02
  0.21311E+07 0.11855E+07-0.17682E+06 0.32324E+04-0.74724E+01
  0.52460E+08-0.50666E+07 0.24593E+06-0.14657E+04 0.29084E-01
 TOTAL ATOMIC SPHERE DENSITY EXPANSION COEFFICIENTS:
  0.37633E+06-0.24236E+06 0.76144E+05-0.12399E+05 0.19560E+04
  0.26653E+07 0.14038E+07-0.24522E+06 0.15607E+05-0.12038E+04
  0.62417E+08-0.76466E+07 0.45470E+06-0.13302E+05 0.20066E+03
```

SLATER-KOSTER INTEGRALS:

HAMILTONIAN INTEGRALS:

ONE-CENTRE:

(S)	(P0)	(P1-)	(P1+)	(D0)	(D1-)	(D1+)	(D2-)	(D2+)
0.1137	0.3252	0.3252	0.3252	0.1301	0.1286	0.1286	0.1286	0.1301

TWO-CENTRE:

(SSS)	(PSS)	(DSS)	(PPS)	(PPP)	(DPS)	(DPP)	(DDS)	(DDP)	(DDD)
-0.0296	-0.0070	-0.0140	-0.0587	-0.0092	0.0096	-0.0114	-0.0125	0.0087	-0.0016
-0.0162	-0.0227	-0.0059	0.0302	-0.0065	0.0090	-0.0024	-0.0026	0.0009	-0.0001
-0.0045	-0.0071	-0.0014	0.0124	-0.0017	0.0023	-0.0005	-0.0005	0.0002	-0.0000
-0.0016	-0.0026	-0.0006	0.0049	-0.0001	0.0011	-0.0000	-0.0002	0.0000	0.0000

OVERLAP INTEGRALS:

ONE-CENTRE:

(S)	(P0)	(P1-)	(P1+)	(D0)	(D1-)	(D1+)	(D2-)	(D2+)
0.9814	0.9782	0.9782	0.9782	0.9995	0.9991	0.9991	0.9991	0.9995

TWO-CENTRE:

(SSS)	(PSS)	(DSS)	(PPS)	(PPP)	(DPS)	(DPP)	(DDS)	(DDP)	(DDD)
0.1843	0.2557	0.0613	-0.3663	0.0948	-0.0965	0.0421	0.0314	-0.0194	0.0030
0.0291	0.0424	0.0097	-0.0763	0.0059	-0.0186	0.0027	0.0044	-0.0011	0.0001
0.0044	0.0036	0.0009	-0.0050	-0.0004	-0.0015	0.0000	0.0003	-0.0000	-0.0000
0.0009	-0.0002	0.0001	0.0016	-0.0005	0.0000	-0.0001	0.0000	0.0000	-0.0000

FERMI ENERGY = 0.1831 HARTREE

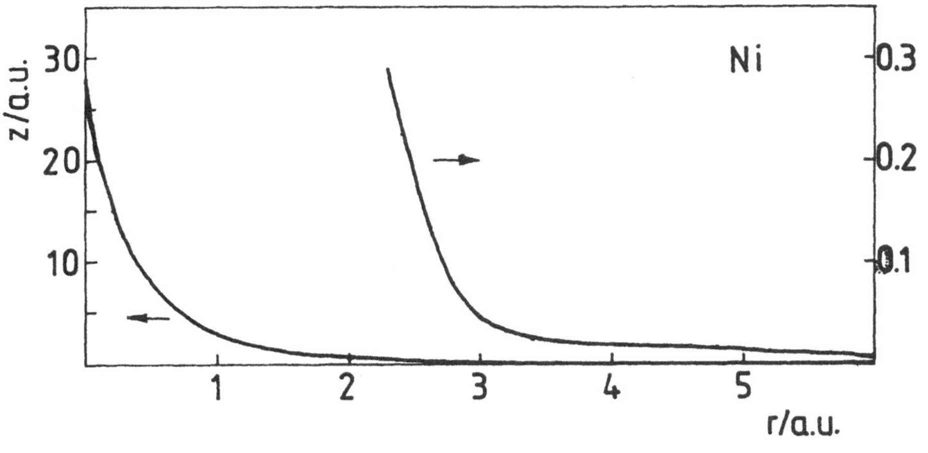

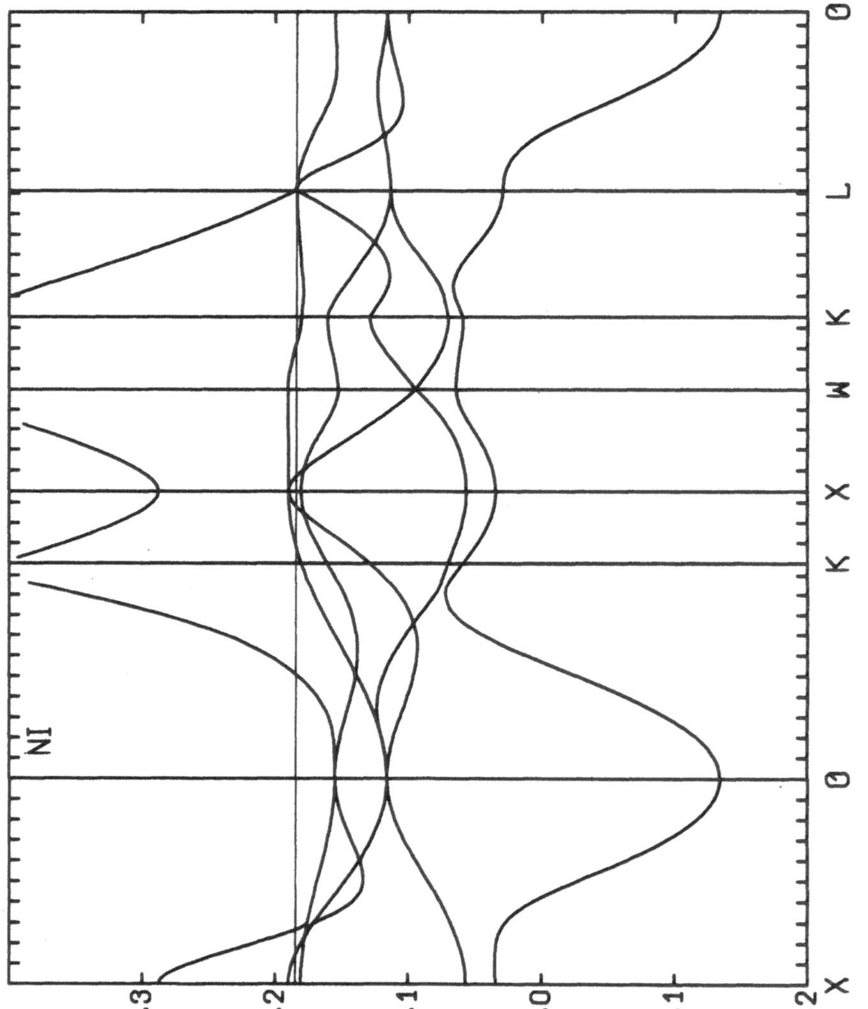
NI
ENERGY/HARTREE
0.3
0.2
0.1
0.0
-0.1
-0.2
X
0
K
X
W
K
L
0

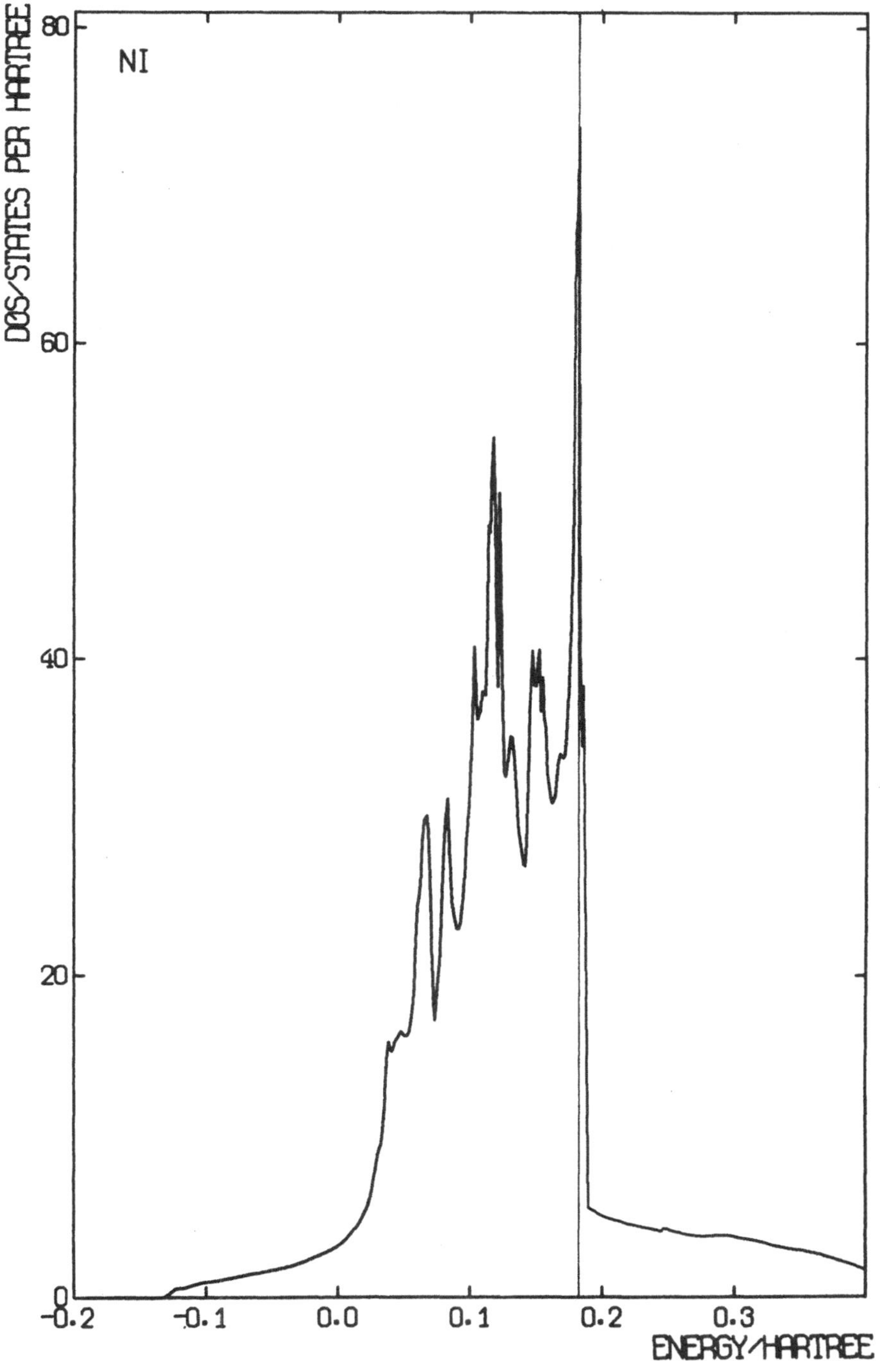
NI
DOS/STATES PER HARTREE
80
60
40
20
0
-0.2
-0.1
0.0
0.1
0.2
0.3
ENERGY/HARTREE

```
COPPER, F.C.C. STRUCTURE, A = 6.82 A.U.

 EXPONENTS FOR THE POTENTIAL BASIS:
 29.0000      11.0198       4.1874       1.5912
 POTENTIAL EXPANSION COEFFICIENTS:
  0.23376E+02-0.14065E+02-0.20891E+01 0.21778E+02
  0.22037E+03 0.15428E+03 0.36445E+02-0.18776E+01
  0.14867E+04-0.63848E+03-0.10150E+03-0.70808E+00

 EXPONENTS FOR THE ATOM-LIKE ORBITAL BASIS:
 29.0000      11.0198       4.1874       1.5912
 ORBITAL BASIS EXPANSION COEFFICIENTS:
   1S:
  0.31867E+03-0.19059E+02 0.46082E+01-0.24173E+00
  0.21170E+03 0.69020E+02-0.76822E+01 0.12922E+00
  0.29276E+04-0.22714E+03 0.49924E+01-0.17200E-01
   2S:
 -0.87811E+02-0.12752E+02 0.74860E+01-0.29499E+00
 -0.52339E+03 0.57992E+03-0.11845E+02 0.15576E+00
 -0.18972E+04-0.44180E+03 0.69591E+01-0.20494E-01
   2P:
 -0.15759E+03-0.55161E+03 0.31841E+01-0.67552E-01
 -0.10837E+04 0.65669E+03-0.49420E+01 0.32709E-01
 -0.50262E+04-0.84106E+03 0.24838E+01-0.39098E-02
   3S:
 -0.32666E+02 0.10808E+01-0.28908E+01-0.13060E+01
 -0.21263E+03 0.33904E+03-0.41322E+02 0.62099E+00
 -0.79249E+03 0.34944E+03-0.29196E+00-0.74805E-01
   3P:
 -0.52076E+02-0.26469E+03 0.49800E+02 0.53374E+00
 -0.37451E+03 0.17127E+02-0.14589E+02-0.21207E+00
 -0.17318E+04-0.36742E+03 0.16242E+02 0.21068E-01
   3D:
 -0.55296E+02 0.24510E+03 0.53991E+02 0.11087E+01
 -0.15295E+03-0.17483E+03-0.28866E+02-0.22140E+00
 -0.75644E+04 0.14011E+04 0.14713E+02 0.68842E-02
   4S:
 -0.56940E+02 0.21373E+03-0.19915E+03 0.52716E+02
 -0.58617E+03 0.18528E+02 0.27137E+03-0.34745E+02
 -0.29202E+04 0.26626E+04-0.37235E+03 0.35064E+01
   4P:
  0.63152E+02-0.24163E+02 0.61010E+02-0.20452E+02
  0.55477E+03 0.84242E+02-0.91846E+02 0.15132E+02
  0.39194E+04-0.12344E+04 0.13259E+03-0.19395E+01

 EXPONENTS FOR THE DENSITY BASIS:
 58.0000      23.6048       9.6067       3.9097       1.5912
 SITE DENSITY EXPANSION COEFFICIENTS:
  0.25134E+06-0.54246E+05 0.10740E+05 0.14294E+03 0.32276E+01
  0.82892E+06 0.36605E+06-0.64486E+05 0.47586E+03 0.64542E+01
  0.16994E+08 0.10334E+07 0.10244E+06-0.28291E+03-0.12369E+01
 TOTAL ATOMIC SPHERE DENSITY EXPANSION COEFFICIENTS:
  0.27511E+06-0.90857E+05 0.30495E+05-0.67365E+04 0.13237E+04
  0.11926E+07 0.52152E+06-0.10914E+06 0.85129E+04-0.82176E+03
  0.24620E+08-0.77248E+06 0.24200E+06-0.82434E+04 0.13845E+03
```

SLATER-KOSTER INTEGRALS:

HAMILTONIAN INTEGRALS:

ONE-CENTRE:

(S)	(P0)	(P1-)	(P1+)	(D0)	(D1-)	(D1+)	(D2-)	(D2+)
0.0954	0.3057	0.3057	0.3057	0.0943	0.0929	0.0929	0.0929	0.0943

TWO-CENTRE:

(SSS)	(PSS)	(DSS)	(PPS)	(PPP)	(DPS)	(DPP)	(DDS)	(DDP)	(DDD)
-0.0365	0.0196	0.0151	-0.0379	-0.0123	0.0140	-0.0114	-0.0110	0.0071	-0.0012
-0.0150	0.0216	0.0048	0.0299	-0.0059	0.0077	-0.0019	-0.0019	0.0007	-0.0001
-0.0039	0.0060	0.0010	0.0110	-0.0013	0.0018	-0.0003	-0.0003	0.0001	-0.0000
-0.0012	0.0019	0.0004	0.0038	-0.0000	0.0007	-0.0000	-0.0001	0.0000	0.0000

OVERLAP INTEGRALS:

ONE-CENTRE:

(S)	(P0)	(P1-)	(P1+)	(D0)	(D1-)	(D1+)	(D2-)	(D2+)
0.9886	0.9856	0.9856	0.9856	0.9998	0.9996	0.9996	0.9996	0.9998

TWO-CENTRE:

(SSS)	(PSS)	(DSS)	(PPS)	(PPP)	(DPS)	(DPP)	(DDS)	(DDP)	(DDD)
0.1724	-0.2446	-0.0515	-0.3569	0.0900	-0.0831	0.0348	0.0238	-0.0139	0.0021
0.0260	-0.0382	-0.0075	-0.0705	0.0053	-0.0147	0.0021	0.0029	-0.0007	0.0000
0.0038	-0.0029	-0.0007	-0.0042	-0.0005	-0.0011	-0.0000	0.0002	-0.0000	-0.0000
0.0007	0.0005	-0.0001	0.0020	-0.0005	0.0001	-0.0001	0.0000	0.0000	-0.0000

FERMI ENERGY = 0.1954 HARTREE

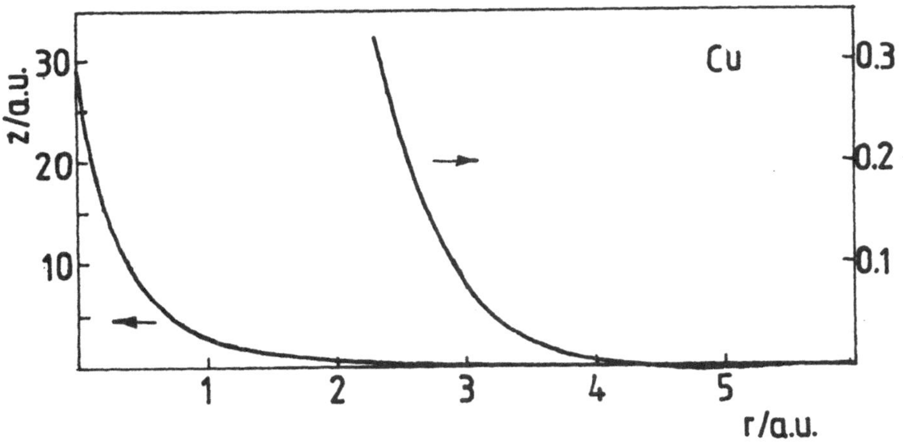

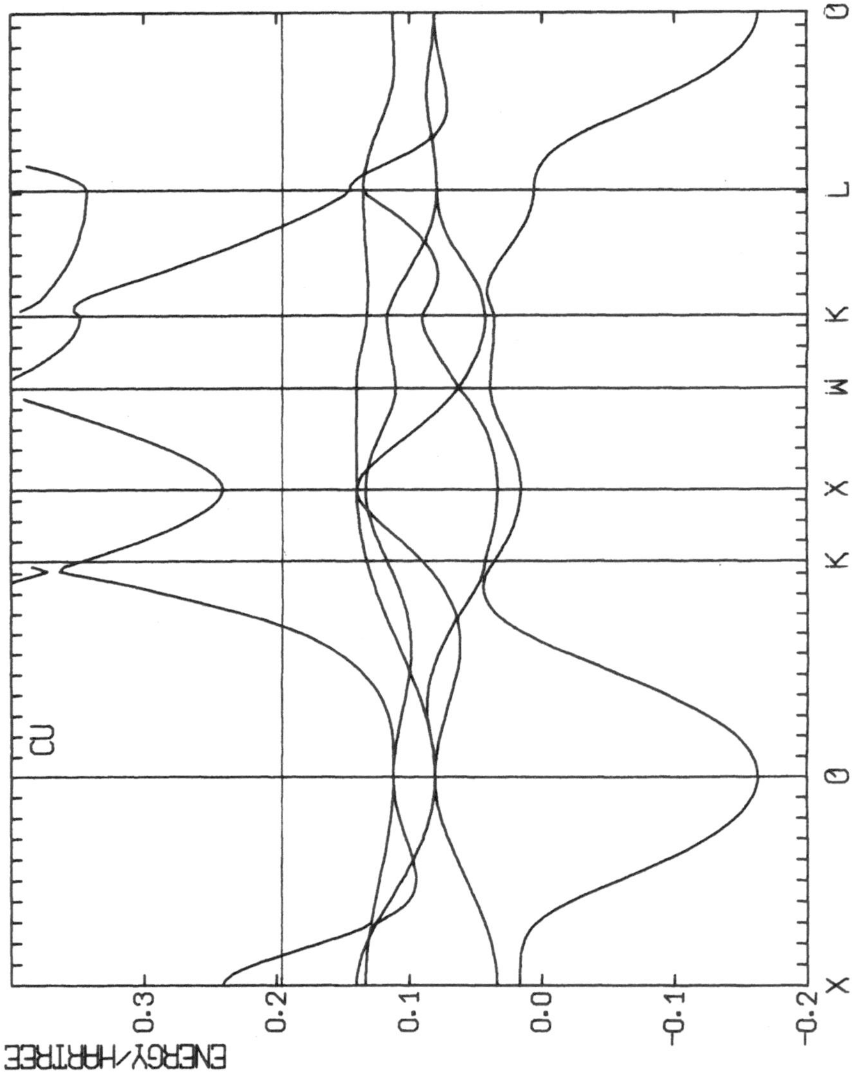
CU
ENERGY/HARTREE
0.3
0.2
0.1
0.0
-0.1
-0.2
X
K
X
W
K
L

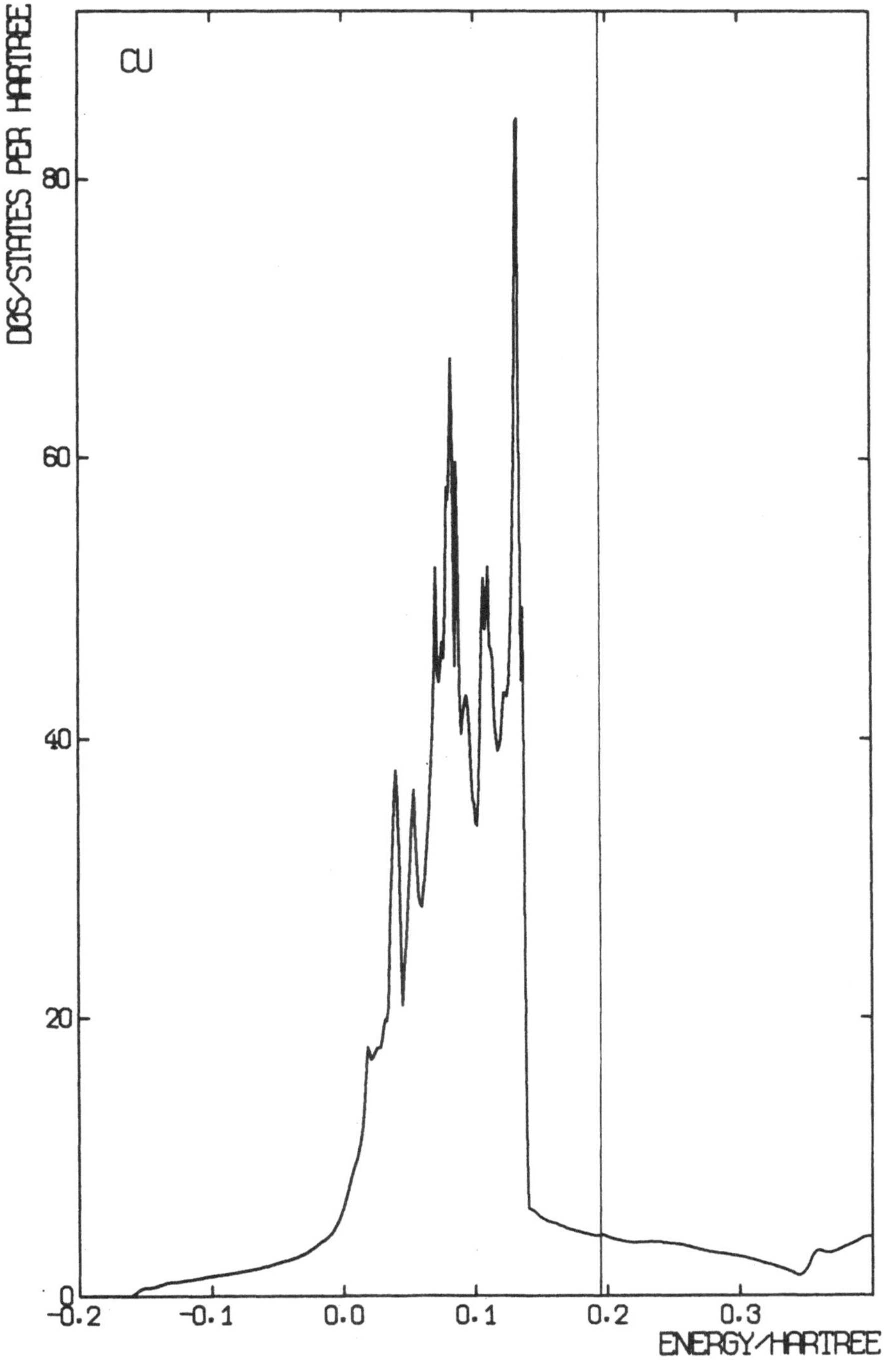
CU
DOS/STATES PER HARTREE
0
20
40
60
80
-0.2
-0.1
0.0
0.1
0.2
0.3
ENERGY/HARTREE

```
ZINK, H.C.P. STRUCTURE, A = 5.03 A.U., C = 9.34 A.U.

 EXPONENTS FOR THE POTENTIAL BASIS:
 30.0000      10.9677      4.0097      1.4659
 POTENTIAL EXPANSION COEFFICIENTS:
 -0.40844E+01 0.32637E+02-0.27541E+02 0.28988E+02
 -0.67276E+02 0.39070E+02 0.58272E+02-0.88667E+01
  0.13808E+03 0.14194E+03-0.13731E+03 0.10806E+01

 EXPONENTS FOR THE ATOM-LIKE ORBITAL BASIS:
 30.0000      10.9677      4.0097      1.4659
 ORBITAL BASIS EXPANSION COEFFICIENTS:
   1S:
 -0.33261E+03 0.16068E+02-0.34955E+01 0.17658E+00
 -0.20236E+03-0.63137E+02 0.58003E+01-0.90883E-01
 -0.32254E+04 0.19715E+03-0.36748E+01 0.11598E-01
   2S:
  0.10003E+03 0.57007E+01-0.72768E+01 0.30029E+00
  0.64973E+03-0.58398E+03 0.11630E+02-0.15330E+00
  0.25773E+04 0.61464E+03-0.68607E+01 0.19414E-01
   2P:
  0.19282E+03 0.58555E+03-0.37740E+01 0.70241E-01
  0.14217E+04-0.87033E+03 0.56993E+01-0.32565E-01
  0.81754E+04 0.95245E+03-0.27326E+01 0.37126E-02
   3S:
 -0.33331E+02 0.44897E+01-0.85027E+01-0.65435E+00
 -0.22403E+03 0.41057E+03-0.35021E+02 0.31714E+00
 -0.89133E+03 0.21327E+03 0.53246E+01-0.38485E-01
   3P:
 -0.66541E+02-0.27870E+03 0.50761E+02 0.97371E-01
 -0.51691E+03 0.10220E+03-0.20531E+02-0.33491E-01
 -0.29512E+04-0.19305E+03 0.13229E+02 0.27954E-02
   3D:
 -0.33011E+02 0.27082E+03 0.55858E+02 0.61876E+00
  0.49300E+02-0.76609E+02-0.36967E+02-0.14294E+00
 -0.62315E+04 0.13645E+04 0.17459E+02 0.74414E-02
   4S:
 -0.26577E+02 0.10959E+03-0.10118E+03 0.28622E+02
 -0.31164E+03-0.72628E+02 0.14247E+03-0.19728E+02
 -0.17951E+04 0.14636E+04-0.18968E+03 0.18654E+01
   4P:
  0.60229E+02-0.57024E+01 0.35821E+02-0.11727E+02
  0.53995E+03 0.99119E+02-0.62374E+02 0.89004E+01
  0.46983E+04-0.10675E+04 0.81918E+02-0.10835E+01

 EXPONENTS FOR THE DENSITY BASIS:
 60.0000      23.7214      9.3784      3.7078      1.4659
 SITE DENSITY EXPANSION COEFFICIENTS:
  0.29974E+06-0.73405E+05 0.73861E+04-0.20908E+03 0.19522E+02
  0.11522E+07 0.71596E+06-0.49065E+05 0.73001E+03-0.38976E+01
  0.30492E+08 0.32607E+06 0.10099E+06-0.41326E+03 0.20626E+00
 TOTAL ATOMIC SPHERE DENSITY EXPANSION COEFFICIENTS:
  0.31715E+06-0.98347E+05 0.19373E+05-0.40977E+04 0.75934E+03
  0.14094E+07 0.84162E+06-0.77908E+05 0.51974E+04-0.44955E+03
  0.37278E+08-0.97192E+06 0.18503E+06-0.46688E+04 0.72310E+02
```

```
SLATER-KOSTER INTEGRALS:

  HAMILTONIAN INTEGRALS:

  ONE-CENTRE:
  ( S )  (PO )  (P1-)  (P1+)  (DO )  (D1-)  (D1+)  (D2-)  (D2+)
 -0.0027 0.2185 0.1943 0.1943-0.1190-0.1184-0.1184-0.1204-0.1204

  TWO-CENTRE:
  (SSS)  (PSS)  (DSS)  (PPS)  (PPP)  (DPS)  (DPP)  (DDS)  (DDP)  (DDD)
 -0.0467 0.0388 0.0168-0.0072-0.0163 0.0210-0.0115-0.0093 0.0051-0.0008
 -0.0373 0.0369 0.0123 0.0103-0.0145 0.0169-0.0076-0.0057 0.0029-0.0004
 -0.0111 0.0170 0.0027 0.0260-0.0044 0.0048-0.0010-0.0008 0.0002-0.0000
 -0.0019 0.0030 0.0004 0.0059-0.0004 0.0006-0.0001-0.0001 0.0000-0.0000
 -0.0038 0.0063 0.0008 0.0120-0.0012 0.0014-0.0002-0.0002 0.0000-0.0000
 -0.0026 0.0043 0.0005 0.0084-0.0007 0.0009-0.0001-0.0001 0.0000-0.0000
 -0.0004 0.0005 0.0001 0.0011 0.0000 0.0001 0.0000-0.0000-0.0000 0.0000
 -0.0010 0.0016 0.0002 0.0035-0.0001 0.0003 0.0000-0.0000 0.0000 0.0000

  OVERLAP INTEGRALS:

  ONE-CENTRE:
  ( S )  (PO )  (P1-)  (P1+)  (DO )  (D1-)  (D1+)  (D2-)  (D2+)
  0.9823 0.9864 0.9775 0.9775 0.9994 0.9998 0.9998 0.9994 0.9994

  TWO-CENTRE:
  (SSS)  (PSS)  (DSS)  (PPS)  (PPP)  (DPS)  (DPP)  (DDS)  (DDP)  (DDD)
  0.1664-0.2429-0.0376-0.3545 0.0978-0.0632 0.0268 0.0143-0.0078 0.0012
  0.1176-0.1793-0.0266-0.2850 0.0625-0.0483 0.0164 0.0093-0.0043 0.0006
  0.0203-0.0305-0.0039-0.0583 0.0055-0.0079 0.0014 0.0010-0.0003 0.0000
  0.0023-0.0020-0.0005-0.0033-0.0011-0.0007-0.0002 0.0001 0.0000-0.0000
  0.0045-0.0054-0.0010-0.0106-0.0013-0.0018-0.0002 0.0003 0.0000-0.0000
  0.0033-0.0034-0.0007-0.0063-0.0011-0.0011-0.0002 0.0002 0.0000-0.0000
 -0.0001 0.0013 0.0001 0.0032-0.0005 0.0003-0.0001-0.0000 0.0000-0.0000
 -0.0007 0.0028 0.0005 0.0058-0.0005 0.0010-0.0001-0.0002 0.0000-0.0000

 FERMI ENERGY =   0.1628 HARTREE
```

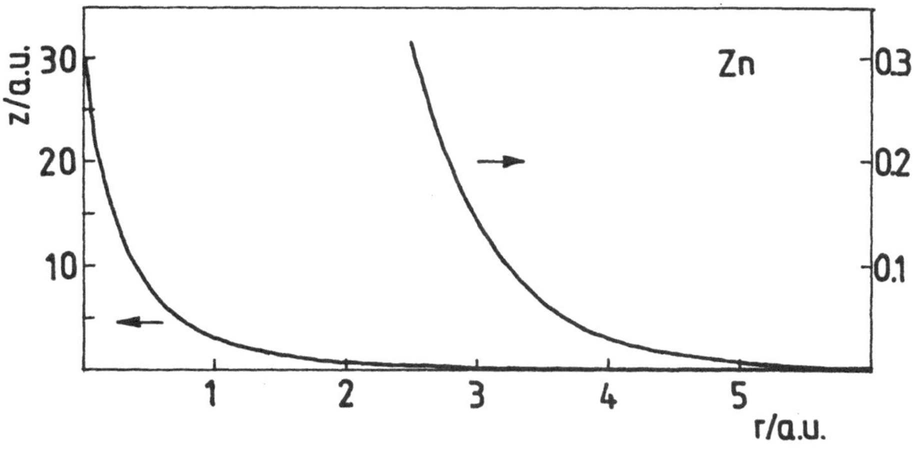

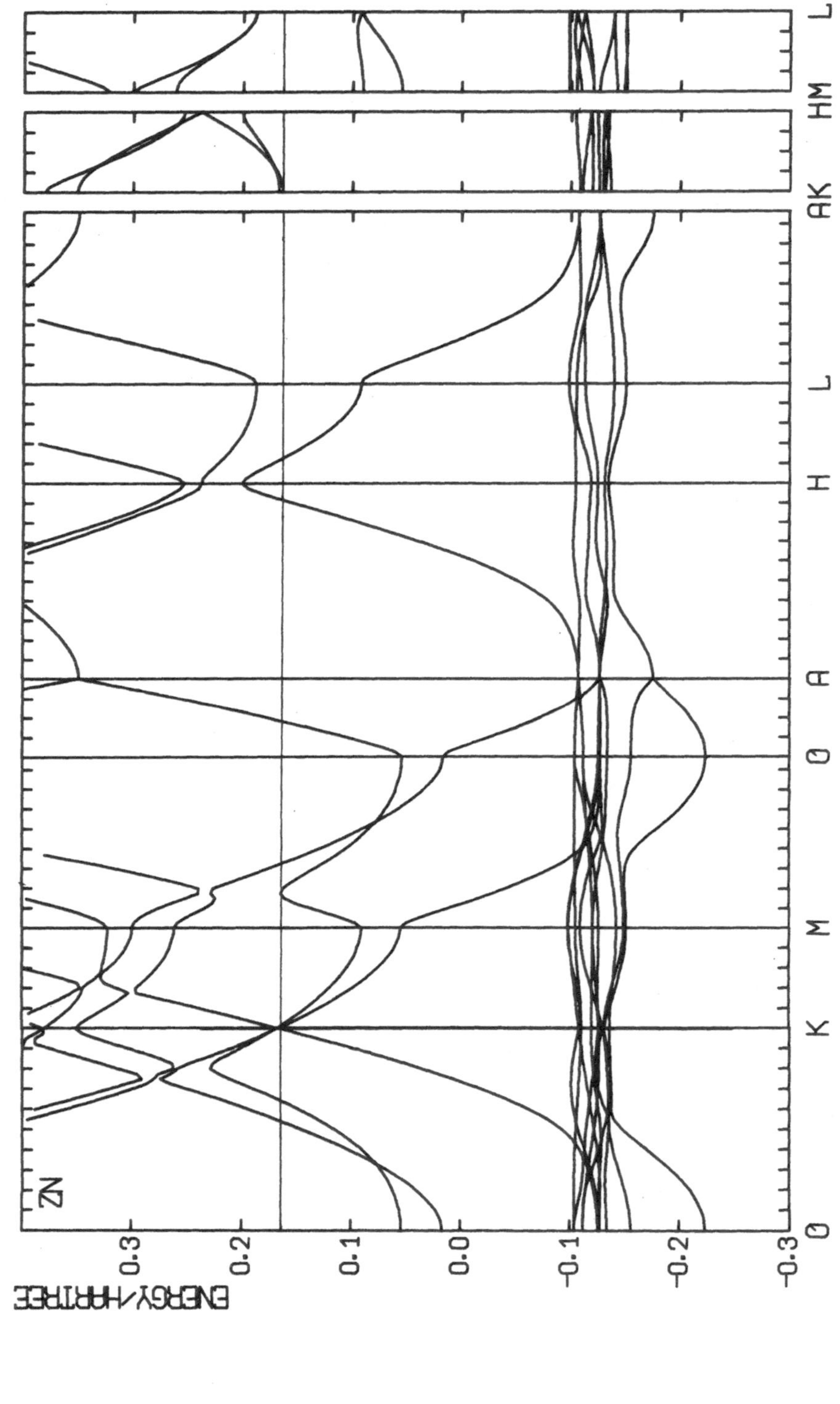
ZN
ENERGY/HARTREE
0.3
0.2
0.1
0.0
-0.1
-0.2
-0.3
Γ
K
M
Q
A
H
L
AK
HM
L

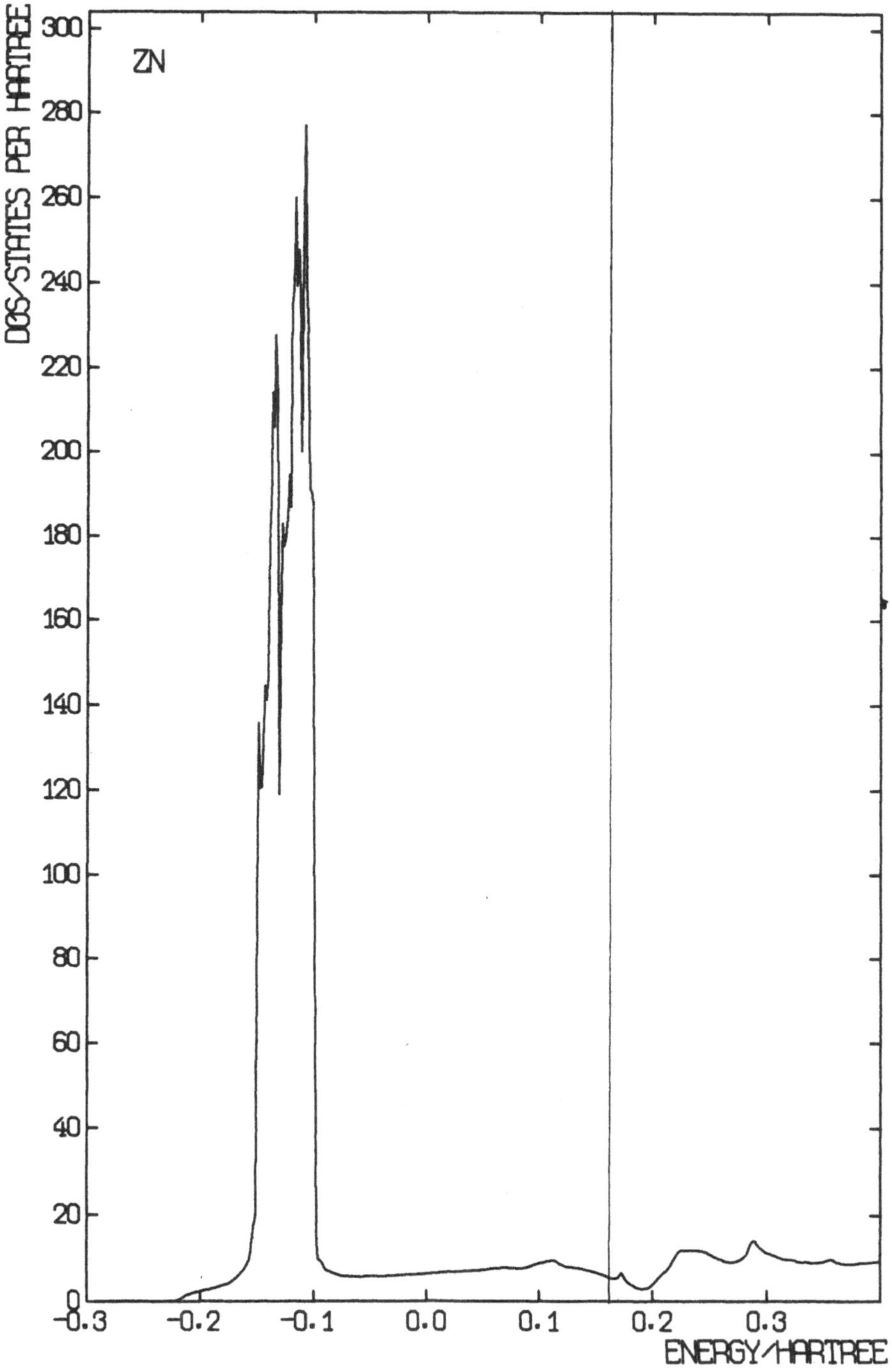
DOS/STATES PER HARTREE
ZN
300
280
260
240
220
200
180
160
140
120
100
80
60
40
20
0
-0.3
-0.2
-0.1
0.0
0.1
0.2
0.3
ENERGY/HARTREE

References

[1] Andersen, O. K., and O. Jepsen, Phys. Rev. Letters **53**, 2571 (1984).

[2] Andersen, O. K., O. Jepsen, and D. Glötzel in: F.Bassani. F. Fumi, and M. P. Tosi (eds.), "Highlights of Condensed-Matter Theory", Proc. Int. School of Theor. Phys. "Enrico Fermi", Course LXXXIX, North-Holland, Amsterdam, 1985.

[3] Anderson, P. W., Phys. Rev. Letters **21**, 13 (1968).

[4] Anderson, P. W., Phys. Rev. **181**, 25 (1969).

[5] Andreoni, W., Phys. Rev. **B14**, 4247 (1976).

[6] Appelbaum, J. A., J. Feibelmann, and D. R. Hamann, Phys. Rev. **B20**, 1433 (1979).

[7] Appelbaum, J. A., and D. R. Hamann in: M. J. G. Lee, J. M. Perz, and E. Fawcett (eds.), "Transition Metals", Institute of Physics, Bristol, 1980.

[8] Bansil, A., Solid State Commun. **16**, 885 (1975).

[9] Bonch-Bruevich, V. L., JETP **31**, 522 (1956) (in Russian).

[10] Born, M. and K. Huang, "Dynamical Theory of Crystal Lattices", Clarendon, Oxford, 1954.

[11] Born, M., and R. Oppenheimer, Ann. Physik (Leipz.) **84**, 457 (1927).

[12] Brauer, W., "Einführung in die Elektronentheorie der Metalle", Geest & Portig, Leipzig, 1972.

[13] Brovman, E. G., and Ju. M. Kagan in: G. K. Horton and A. A. Maradudin (eds.), "Dynamical Properties of Solids", vol. 1, p. 191, North-Holland, Amsterdam, 1974.

[14] Brust, D., J. Comput. Phys. **8**, 33 (1968).

[15] Bullett, D. W., Solid State Physics **35**, 129 (1980).

[16] Burdick, G. A., Phys. Rev. **129**, 138 (1963).

[17] Buttler, A. L., S. S. Ipson, and A. H. Lettington, J. Phys. **C16**, 6335 (1983).

[18] Cabannes, H. (ed.), "Padé Approximants Method and its Applications to Mechanics", Lecture Notes in Physics **47**, Springer, Berlin, 1976.

[19] Callaway, J., and A. J. Hughes, Phys. Rev. **156**, 860 (1967).

[20] Callaway, J., D. Laurent, and C. S. Wang, Inst. Phys. Conf. Ser. **39**, 41 (1978).

[21] Callaway, J., and H. M. Zhang, Phys. Rev. **B1**, 305 (1970).

[22] Chadi, D. J., and M. L. Cohen, Solid State Commun. **13**, 1007 (1973).

[23] Clinton, W. L., and B. Rice, J. Chem. Phys. **30**, 542 (1959).

[24] Cohen, M. L., Physica Scripta **1**, 5 (1982).

[25] Des Cloizeaux, J., Phys. Rev. **129**, 554 (1963).

[26] Des Cloizeaux, J., Phys. Rev. **135**, A685; A698 (1964).

[27] Dederichs, P. H., and R. Zeller, Festkörperprobleme/ Advances in Solid State Physics (Pergamon/ Vieweg) **XXI**, 243 (1981).

[28] Dederichs, P. H., and R. Zeller in: P. Ziesche (ed.), Proc. 12th Symp. Electronic Structure, p.75, TU Dresden, 1982.

[29] Elliott, R. J., Physica **86–88B**, 1118 (1977).

[30] Englisch, H., and R. Englisch, phys. stat. sol. (b)**123**, 711 (1984).

[31] Eschrig, H. in: P. Ziesche (ed.), Proc. 4th Symp. Electronic Structure, p. 17, TU Dresden, 1974.

[32] Eschrig, H., phys. stat. sol. (b)**96**, 329 (1979).

[33] Eschrig, H. in: P. Ziesche (ed.), Proc. 9th Symp. Electronic Structure, p. 17, TU Dresden, 1979.

[34] Eschrig, H. in: P. Ziesche (ed.), Proc. 16th Symp. Electronic Structure, p. 49, TU Dresden, 1986.

[35] Eschrig, H., and I. Bergert, phys. stat. sol. (b)**90**, 621 (1978).

[36] Eschrig, H., K. A. Kikoin, and V. G. Kohn, Solid St. Commun. **56**, 773 (1985).

[37] Eschrig, H., R. Richter, and B. Velicky, J. Phys. **C19**, 7173 (1986).

[38] Eschrig, H., G.Seifert, and P. Ziesche, Solid St. Commun. **56**, 777 (1985).

[39] Ewald, P. P., Ann. Phys. (Leipzig) **64**, 253 (1921).

[40] Feibelmann, P. J., J. A. Appelbaum, and D. R. Hamann, Phys. Rev. **B20**, 1433 (1979).

[41] Filter, E., and E. O. Steinborn, Phys. Rev. **A18**, 1 (1978).

[42] Fröhlich, H., Proc. Roy. Soc. **A223**, 296 (1954).

[43] Fulde, P. in: K. A. Gschneidner and L. Eyring (eds.) "Handbook on the Physics and Chemistry of Rare Earths", North-Holland, Amsterdam, 1978.

[44] Galitzky, V. M., and A. B. Migdal, JETP **34**, 151; 1011 (1958) (in Russian).

[45] Gell-Mann, M., and F. Low, Phys. Rev. **84**, 350 (1951).

[46] Gilat, G., Methods Comp. Phys. **15**, 317 (1976).

[47] Gilat, G., and G. Dolling, Phys. Letters **8**, 304 (1964).

[48] Gilat, G., and L. J. Raubenheimer, Phys. Rev. **144**, 390 (1966).

[49] Glötzel, D., B. Segall, and O. K. Andersen, Solid St. Commun. **36**, 403 (1980).

[50] Griepentrog, E., and A. Möbius, Z. angew. Math. Mech. **65**, 561 (1985).

[51] Gunnarsson, O., M. Jonson, and B. I. Lundqvist, Phys. Rev. **B20**, 3136 (1979).

[52] Guseinov, I. I., J. Phys. **B3**, 1399 (1970).

[53] Haydock, R., V. Heine, and M. J. Kelly, J. Phys. **C8**, 2591 (1975).

[54] Haydock, R., Solid State Physics **35**, 216 (1980).

[55] Haydock, R., and C. M. M. Nex, J. Phys. **C18**, 2235 (1985).

[56] Harrison, W. A., "Solid State Theory", chap. II.7 and chap. IV.4, McGraw-Hill book company, New York, 1970.

[57] Harrison, W. A., "Electronic Structure and the Properties of Solids", Freeman and Company, San Francisco, 1980.

[58] Hedin, L., and S. Lundqvist, Solid State Physics **23**, 1 (1969).

[59] Hedin, L., and B. I. Lundqvist, J. Phys. **C4**, 2064 (1971).
[60] Hohenberg, P., and W. Kohn, Phys. Rev. **136**, B864 (1964).
[61] Hoshino, T., T. Asada, and K. Terakura, Phys. Rev. **B31**, 2005 (1985)
[62] Hybertsen, M. S., and S. G. Louie, Phys. Rev. Letters **55**, 1418 (1985); Phys. Rev. **B32**, 7005 (1985).
[63] Jahn, H. A., Proc. Roy. Soc. **A161**, 117 (1938).
[64] Jahn, H. A. and E. Teller, Proc. Roy. Soc. **A161**, 220 (1937).
[65] Jepsen, O., and O. K. Andersen, Solid State Commun. **9**, 1763 (1971).
[66] Jones, H., "The Theory of Brillouin Zones and Electronic States in Crystals", North-Holland, Amsterdam, and American Elsevier, New York, 1975.
[67] Jones, R. O., J. Chem. Phys. **76**, 2098 (1982).
[68] Kane, E. O., and A. B. Kane, Phys. Rev. **B17**, 2691 (1978).
[69] Keller, J., J. Phys. **C4**, L85 (1971).
[70] Kelly, M. J., Solid State Physics **35**, 296 (1980).
[71] Kleinman, L., Phys. Rev. **B28**, 1139 (1983).
[72] Klima, J., J. Phys. **C12**, 3691 (1979).
[73] Kohn, W., Phys. Rev. **115**, 809 (1959).
[74] Kohn, W., Phys. Rev. **B7**, 4388 (1973).
[75] Kohn, W., and L. J. Sham, Phys. Rev. **140**, A1133 (1965).
[76] Koster, G. F., Phys. Rev. **89**, 67 (1953).
[77] Krüger, E., phys. stat. sol. (b)**52**, 215, 512 (1972).
[78] Kudrnovsky, J., and B. Velicky in: P. Ziesche (ed.), Proc. 9-th Symp. Electronic Structure, p. 168, TU Dresden, 1979.
[79] Kunc, K., Helvetica Phys. Acta **56**, 559 (1983).
[80] Kunc, K., in: J. T. Devreese and P. E. Van Camp (eds.), "Electronic Structure, Dynamics and Quantum Structural Properties of Condensed Matter", Plenum Press, New York, 1985, p. 227.
[81] Lafon, E. E. and C. C. Lin, Phys. Rev. **152**, 579 (1966).
[82] Landau, L. D., JETP **11**, 592 (1941) (in Russian).
[83] Landau, L. D., JETP **30**, 1058 (1956) (in Russian).
[84] Landau, L. D., V. B. Berestetskii, E. M. Lifshitz, and L. P. Pitaevskii, "Statistical Physics" (part I), Pergamon Press, London, 1980.
[85] Landau, L. D., and E. M. Lifshitz, "Quantum Mechanics", Pergamon Press, London, 1977.
[86] Landau, L. D., and E. M. Lifshitz, and L. P. Pitaevskii, "Statistical Physics" (part II), Pergamon Press, London, 1980.
[87] Landolt-Börnstein, Numerical Data and Functional Relations in Science and Technology, New Series, vol. III/6, Springer, Heidelberg, 1971.
[88] Lehmann, G., P. Rennert, M. Taut, and H. Wonn, phys. stat. sol. **37**, K27 (1970).
[89] Lehmann, G., and M. Taut, phys. stat. sol. (b)**54**, 469 (1972); **57**, 815 (1973).
[90] Lieb, E. H., Int. J. Quantum Chem. XXIV, 243 (1983).

[91] Lix, B., phys. stat. sol. (b)**44**, 411, (1971).
[92] Löwdin, P. O., J. Chem. Phys. **18**, 365 (1950).
[93] Luttinger, J. M., Phys. Rev. **119**, 1153 (1960).
[94] Maradudin, A. A. in: G. K. Horton and A. A. Maradudin (eds.), "Dynamical Properties of Solids", vol. 1, p. 1, North-Holland, Amsterdam, 1974.
[95] Mertig, I., and E. Mrosan, phys. stat. sol. (b)**94**, K23 (1979).
[96] Moruzzi, V. L., J. F. Janak, and A. R. Williams, Calculated Electronic Properties of Metals, Pergamon Press, New York, 1978.
[97] Mueller, F. M., J. W. Garland, M. H. Cohen, and K. H. Bennemann, Ann. Phys. (USA) **67**, 19 (1971).
[98] Müller, J. E., R. O. Jones, and J. Harris, J. Chem. Phys. **79**, 1875 (1983).
[99] Nuttall, J., Phys. Rev. **157**, 1312 (1967).
[100] Papaconstantopoulos, D. A., "Handbook of Band Structure of Elemental Solids", Plenum Press, New York, 1986.
[101] Parzen, G., Phys. Rev. **89**, 237 (1953).
[102] Perdew, J. P., and M. Levy, Phys. Rev. Lett. **51**, 1884 (1983).
[103] Phillips, J. C., and L. Kleinman, Phys. Rev. **116**, 287 (1959).
[104] Podloucky, R., R. Zeller, and P. H. Dederichs, Phys. Rev. **B22**, 5777 (1980).
[105] Prasad, R., and A. Bansil, Phys. Rev. **B21**, 496 (1980).
[106] Richter, M., and H. Eschrig in: P. Ziesche (ed.), Proc. 16th Symp. Electronic Structure, p. 132, TU Dresden, 1986.
[107] Richter, R., H. Eschrig, and B. Velicky, J. Phys. **F17**, 351 (1987).
[108] Roothaan, C. C. J., J. Chem. Phys. **19**, 1445 (1951).
[109] Sambe, H., and R. H. Felton, J. Chem. Phys. **62**, 1122 (1975).
[110] Schwarz, K., Phys. Rev. **B5**, 2466 (1972).
[111] Schwarz, K., Theor. Chim. Acta (Berlin) **34**, 225 (1974).
[112] Sham, L. J., and M. Schlüter, Phys. Rev. Lett. **51**, 1888 (1983).
[113] Slater, J. C., and G. F. Koster, Phys. Rev. **94**, 1498 (1954).
[114] Taut, M., and H. Eschrig, phys. stat. sol. (b) **73**, 151 (1976).
[115] Taut, M., H. Eschrig, and U.-H. Gläser, J. Phys. **C17**, 3019 (1984).
[116] Tyler, J. M., T. E. Norwood, and J. L. Fry, Phys. Rev. **B1**, 297 (1970).
[117] Weber, W. in: P. Phariseau and W. Temmerman (eds.), "Electronic Structure of Complex Systems", Plenum Press, New York, 1984.
[118] Weger, M., and I. B. Goldberg, Solid State Physics **28**, 1 (1973).
[119] Wilkinson, J. H., "The Algebraic Eigenvalue Problem", Clarendon Press, Oxford, 1965.
[120] Ziesche, P., and G. Lehmann (eds.), "Ergebnisse in der Elektronentheorie der Metalle", Akademie-Verlag, Berlin, 1983.
[121] Ziman, J. M., "Elements of Advanced Quantum Theory", Cambridge, Univ. Press, 1969.
[122] Ziman, J. M., "Principles of the theory of solids", Cambridge University Press, London, 1972.

Subject index

(Terms appearing in the chapter and section headings are not repeated in this subject index list)